GEOProcessing 2023

The Fifteenth International Conference on Advanced Geographic Information Systems, Applications, and Services

April 24th – 28th, 2023

Venice, Italy

GEOProcessing 2023 Editors

Claus-Peter Rückemann, Westfälische Wilhelms-Universität Münster (WWU) / DIMF / Leibniz Universität Hannover, Germany
Rouhollah Nasirzadehdizaji, Water and Environment Department, Yüksel Proje Inc., Turkey

Printed from e-media with permission by:

Curran Associates, Inc.
57 Morehouse Lane
Red Hook, NY 12571

Some format issues inherent in the e-media version may also appear in this print version.

Printed with permission by Curran Associates, Inc. (2024)

International Academy, Research, and Industry Association (IARIA)
412 Derby Way
Wilmington, DE 19810

Phone: (408) 893-6407
Fax: (408) 527-6351

petre@iaria.org

Additional copies of this publication are available from:

Curran Associates, Inc.
57 Morehouse Lane
Red Hook, NY 12571 USA
Phone: 845-758-0400
Fax: 845-758-2633
Email: curran@proceedings.com
Web: www.proceedings.com

Table of Contents

GEOProcessing 2023

Forward

The Fifteenth International Conference on Advanced Geographic Information Systems, Applications, and Services (GEOProcessing 2023), held in Venice, Italy, April 24 - 28, 2023, addressed the aspects of managing geographical information and web services.

The goal of the GEOProcessing 2023 conference was to bring together researchers from the academia and practitioners from the industry in order to address fundamentals of advances in geographic information systems and the new applications related to them using the Web Services. Such systems can be used for assessment, modeling and prognosis of emergencies

GEOProcessing 2023 provided a forum where researchers were able to present recent research results and new research problems and directions related to them. The topics covered aspects from fundamentals to more specialized topics such as 2D & 3D information visualization, web services and geospatial systems, geoinformation processing, and spatial data infrastructure.

We take this opportunity to thank all the members of the GEOProcessing 2023 Technical Program Committee as well as the numerous reviewers. The creation of such a broad and high-quality conference program would not have been possible without their involvement. We also kindly thank all the authors who dedicated much of their time and efforts to contribute to the GEOProcessing 2023. We truly believe that, thanks to all these efforts, the final conference program consists of top quality contributions.

This event could also not have been a reality without the support of many individuals, organizations, and sponsors. We are grateful to the members of the GEOProcessing 2023 organizing committee for their help in handling the logistics and for their work to make this professional meeting a success.

We hope that GEOProcessing 2023 was a successful international forum for the exchange of ideas and results between academia and industry and for the promotion of progress in geographic information research. We also hope that Venice provided a pleasant environment during the conference and everyone saved some time for exploring this beautiful city

GEOProcessing 2023 Chairs

GEOProcessing 2023 General Chair

Claus-Peter Rückemann, Westfälische Wilhelms-Universität Münster (WWU) / DIMF / Leibniz Universität Hannover, Germany

GEOProcessing Steering Committee

Thomas Ritz, FH Aachen, Germany
Yerach Doytsher, Technion - Israel Institute of Technology, Haifa, Israel
Alexey Cheptsov, Information Service Center of the University of Stuttgart (TIK)
Roger Tilley, Sandia National Laboratories, USA
Jui-Hsin (Larry) Lai, Ping-An Technology - Research Lab, USA

GEOProcessing 2023 Publicity Chair

Laura Garcia, Universitat Politècnica de València (UPV), Spain
Javier Rocher Morant, Universitat Politecnica de Valencia, Spain

GEOProcessing 2023

COMMITTEE

GEOProcessing 2023 General Chair

Claus-Peter Rückemann, Westfälische Wilhelms-Universität Münster (WWU) / DIMF / Leibniz Universität Hannover

GEOProcessing Steering Committee

Thomas Ritz, FH Aachen, Germany
Yerach Doytsher, Technion - Israel Institute of Technology, Haifa, Israel
Alexey Cheptsov, Information Service Center of the University of Stuttgart (TIK), Germany
Roger Tilley, Sandia National Laboratories, USA
Jui-Hsin (Larry) Lai, Ping-An Technology - Research Lab, USA

GEOProcessing 2023 Publicity Chairs

Javier Rocher Morant, Universitat Politecnica de Valencia, Spain
Laura Garcia, Universitat Politecnica de Valencia, Spain

GEOProcessing 2023 Technical Program Committee

Mahdi Abdelguerfi, University of New Orleans, USA
Alia I. Abdelmoty, Cardiff University, Wales, UK
Danial Aghajarian, Georgia State University, USA
Nuhcan Akçit, Middle East Technical University, Turkey
Zaher Al Aghbari, University of Sharjah, UAE
Bharath H. Aithal, IIT Kharagpur, India
Md Mahbub Alam, Dalhousie University, Canada
Heba Aly, University of Maryland, College Park, USA
Francisco Javier Ariza López, Escuela Politécnica de Jaén - Universidad de Jaén, Spain
Thierry Badard, Centre de Recherche en Géomatique (CRG) | Université Laval, Canada
Abderrazak Bannari, Arabian Gulf University, Bahrain
Fabian Barbato, Ort University of Uruguay, Uruguay
Melih Basaraner, Yildiz Technical University, Turkey
Peter Baumann, rasdaman GmbH Bremen / Jacobs University Bremen, Germany
Lorenzo Carnevale, University of Messina, Italy
Mete Celik, Erciyes University, Turkey
Aizaz Chaudhry, Carleton University, Canada
Wei Chen, University of Birmingham, UK
Dickson K.W. Chiu, The University of Hong Kong, Hong Kong
Alexey Cheptsov, Information Service Center of the University of Stuttgart (TIK), Germany
Daniel Enrique Constantino Recillas, TESE, Mexico
Alexandre Corrêa da Silva, HEX Geospatial Technologies, Brazil
Mehmet Ali Çullu, Harran University, Turkey
Giacomo De Carolis, CNR-IREA, Italy

Monica De Martino, CNR-IMATI (National research Council, Institute of applied Mathematics and Information technology), Italy
Cláudio de Souza Baptista, University of Campina Grande, Brazil
Subhadip Dey, Indian Institute of Technology Bombay, India
Yerach Doytsher, Technion - Israel Institute of Technology, Haifa, Israel
Suzana Dragicevic, Simon Fraser University, Canada
Surya Durbha, IIT Bombay, India
Emre Eftelioglu, Cargill Inc., USA
Süleyman Eken, Kocaeli University, Turkey
Salah Er-Raki, Université Cadi Ayyad, Morocco
Javier Estornell, Universitat Politècnica de València, Spain
Jamal Ezzahar, Université Cadi Ayyad, Morocco
Zhiwen Fan, University of Texas at Austin, USA
Francisco R. Feito, University of Jaén, Spain
Anabella Ferral, Instituto de Altos Estudios Espaciales Mario Gulich | Centro Espacial Teófilo Tabanera - CONAE, Córdoba, Argentina
Kaiqun Fu, South Dakota State University, USA
Erica Goto, University of Michigan, USA
Mohd Helmy Abd Wahab, Universiti Tun Hussein Onm Malaysia, Malaysia
Arif Hidayat, Monash University, Australia / Brawijaya University, Indonesia
Masaharu Hirota, Okayama University of Science, Japan
Qunying Huang, University of Wisconsin, Madison, USA
Chih-Cheng Hung, Kennesaw State University - Marietta Campus, USA
Sergio Ilarri, University of Zaragoza, Spain
Ge-Peng Ji, Wuhan University (WHU) & Inception Institute of Artificial Intelligence (IIAI), China
Jehn-Ruey Jiang, National Central University, Taiwan
Joaquim João Sousa, INESC TEC, Portugal
Katerina Kabassi, Ionian University, Greece
Hassan A. Karimi, University of Pittsburgh, USA
Baris M. Kazar, Oracle America Inc., USA
Saïd Khabba, Université Cadi Ayyad, Marrakech, Morocco
Kyoung-Sook Kim, National Institute of Advanced Industrial Science and Technology (AIST), Tokyo, Japan
Mel Krokos, University of Portsmouth, UK
Anoop Kumar Shukla, Manipal Academy of Higher Education, India
Piyush Kumar, Florida State University, USA
Jui-Hsin (Larry) Lai, Ping-An Technology - Research Lab, USA
Robert Laurini, INSA Lyon | University of Lyon, France
Dan Lee, Esri Inc.,USA
Lassi Lehto, Finnish Geospatial Research Institute, Finland
Jonathan Li, University of Waterloo, Canada
Jugurta Lisboa-Filho, Federal University of Viçosa, Brazil
Ying Lu, DiDi Research America, Mountain View, USA
Dandan Ma, Northwestern Polytechnic University, China
Ali Mansourian, Lund University, Sweden
Jesús Martí Gavilá, Research Institute for Integrated Management of Coastal Areas (IGIC) | Universitat Politècnica de València, Spain
Sara Migliorini, University of Verona, Italy
Sobhan Moosavi, The Ohio State University, USA

Tathagata Mukherjee, The University of Alabama in Huntsville, USA
Beniamino Murgante, University of Basilicata, Italy
Ahmed Mustafa, The New School University, New York, USA
Purevtseren Myagmartseren, National University of Mongolia, Mongolia
Aldo Napoli, MINES ParisTech - CRC, France
Maurizio Napolitano, Fondazione Bruno Kessler, Trento, Italy
Rouhollah Nasirzadehdizaji, Istanbul University - Cerrahpaşa, Turkey
Alexey Noskov, Philipps University of Marburg, Germany
Xiao Pan, Shijiazhuang Tiedao University, China
Shray Pathak, School of Geographic Sciences | East China Normal University, Shanghai, China
Akshay Patil, AiDASH.inc, India
Kostas Patroumpas, Athena Research Center, Greece
Davod Poreh, Università degli Studi di Napoli "Federico II", Italy
Viktor Prasanna, University of Southern California, USA
Satish Puri, Marquette University, Wisconsin, USA
Kuldeep Purohit, Michigan State University, USA
Honggang Qi, University of Chinese Academy of Sciences, China
Thomas Ritz, FH Aachen, Germany
Ricardo Rodrigues Ciferri, Federal University of São Carlos (UFSCar), Brazil
Armanda Rodrigues, Universidade NOVA de Lisboa | NOVA LINCS, Portugal
Claus-Peter Rückemann, Westfälische Wilhelms-Universität Münster (WWU) / DIMF / Leibniz Universität Hannover, Germany, Germany
Ibrahim Sabek, University of Minnesota, USA
André Sabino, Universidade Autónoma de Lisboa, Portugal
Markus Schneider, University of Florida, USA
Raja Sengupta, McGill University, Montreal, Canada
Elif Sertel, Istanbul Technical University, Turkey
Shih-Lung Shaw, University of Tennessee, Knoxville, USA
Yosio E. Shimabukuro, Brazilian Institute for Space Research - INPE, Brazil
Rajat Shinde, Indian Institute of Technology Bombay, India
Dericks Praise Shukla, IIT Mandi, India
Spiros Skiadopoulos, University of the Peloponnese, Greece
Dimitris Skoutas, Athena Research Center, Greece
Srushti Rashmi Shirish, Woven planet holdings, Tokyo, Japan
Francesco Soldovieri, Istituto per il Rilevamento Elettromagnetico dell'Ambiente - Consiglio Nazionale delle Ricerche (CNR), Italy
Katia Stankov, University of British Columbia, Canada
Ergin Tari, Istanbul Technical University, Turkey
Brittany Terese Fasy, Montana State University, USA
Roger Tilley, Sandia National Laboratories, USA
Goce Trajcevski, Iowa State University, USA
Linh Truong-Hong, Delft University of Technology, Netherlands
Taketoshi Ushiama, Kyushu University, Japan
Marlène Villanova-Oliver, Univ. Grenoble Alpes - Grenoble Informatics Lab, France
Massimo Villari, University of Messina, Italy
Tin Vu, Microsoft Corporation, USA
Hong Wei, University of Maryland, College Park, USA
John P. Wilson, University of Southern California, USA

Copyright Information

Application of InSAR Method to Estimate the Surface Deformation of the May Embankment Dam, Turkey

Rouhollah Nasirzadehdizaji
Department of Water and Environment
Yüksel Proje Inc.
Ankara, 06610, Turkey
e-mail: rnasirzadeh@yukselproje.com.tr

Anil Olgac
Department of Water and Environment
Yüksel Proje Inc.
Ankara, 06610, Turkey
e-mail: aolgac@yukselproje.com.tr

***Abstract*—This study is conducted to apply the Interferometric Synthetic Aperture Radar (InSAR) method to investigate the deformation of the upstream face of the May Embankment Dam located in central Anatolia, Turkey. Optical satellite imagery revealed significant water volume variation at the May Dam over the past decade, prompting investigation due to the high seismic risk of the region and the potential consequences of dam failure for downstream lives and properties. In this study, dam surface displacements were obtained from a series of Sentinel-1 processed Synthetic Aperture Radar (SAR) data through postprocessing and analysis of Single-Look Complex (SLC) radar image pairs (interferograms). The results show that the average displacement of the dam surface is -9.5 cm in the processed 7-year SAR data. However, a cumulative displacement rate of -22.16 cm is estimated in the dam's structure between 2015 and 2022. The highest changes occurred between 2018-2020, with a deformation of -13.37 cm, as the amount of water behind the dam began to increase due to increased precipitation. Analysis of meteorological data indicates that the climate conditions in the semi-arid upstream area that supplies water to the dam result in lower precipitation and reduced water flow for much of the year, often leaving the dam dry or with reduced water volume. However, due to the effects of climate change, heavy precipitation has occurred in the area, which led to large amounts of water collecting behind the dam. Consequently, if not enough attention is paid to the surface changes of the dam that are happening, which have been obtained through the radar satellite data analysis in this study, then during the intensive rainfalls that have already been recorded in the region, there is a potential and possibility of damage to the dam and causing a threat to the lives of the residents of several cities and damage to the agricultural fields on the downstream of the dam.**

Keywords*—*Interferometric SAR; monitoring; surface displacement; dam structure; climate change.

I. Introduction

Dams are one of the most important engineering structures used for water supply, flood control, agriculture, drinking water and hydroelectric power generation. Various parameters play a role in determining the load of the dam. Deformation can occur both in the dam and the surrounding areas. Normally, the structure of the dam, the weight of the embankment and water, water pressure, temperature changes, and the movement of the earth's crust are the deformation factors of a dam [1]. Monitoring is an important part of dam construction and during operation and must be able to detect any behavior that may deteriorate the dam in a timely manner, which may lead to the closure or failure of the dam. Hence, accurate monitoring measurements of the dam and its surrounding areas are required on a regular basis.

Different methods including geodetic and non-geodetic monitoring measurements are applied for dams' deformation controlling and detection. Dam deformation monitoring models can be divided into statistical models, deterministic models, and hybrid models. The statistical model can effectively simulate the functional relationship between dam deformation and environmental variables, such as reservoir water level and temperature [2]. The fractal interpolation function method [3] [4] is applied with the dam prototype observation data processing to predict dam deformation.

Among the monitoring measurement techniques, special emphasis is placed on geodetic methods, and some new measurement techniques are recommended. Static monitoring of the dam based on a geodetic control network is performed by [5] and [6] to measure absolute and relative displacements of the dam structure and the nearby areas, and the geodetic networks were complemented by geotechnical/structural sensors. Geodetic measurements provide information about absolute motion relative to some fixed datum that is considered stable and least susceptible to movement due to the filling and discharging of the dam reservoir. However, the accuracy of the geodetic measurements may be inadequate and costly sometimes when a high frequency of repeated observations is needed. While the precision is achievable, the associated operations are complex and require a team of expert surveyors to work for several days on each campaign. Additionally, terrestrial geodetic surveys require intervisibility between survey points, which limits their application. Geotechnical instruments can be easily adapted for continuous monitoring and, if properly calibrated and installed, can provide highly accurate results. Though, these instruments only provide local information on selected parts of the deformation.

Over the past two decades, innovative remote sensing techniques based on satellite SAR data (satellites that acquire data using the microwave portion of the electromagnetic spectrum) have been developed to detect and monitor surface displacements. InSAR is a radar technique used in geodesy and remote sensing to generate maps of surface deformation and potentially measure sub-centimeter scale deformation changes over days to years. The Differential InSAR (DInSAR),

Advanced stacking DInSAR (A-DInSAR), Multi-Temporal InSAR (MT-InSAR), production of accurate velocity maps and displacement time-series methods are among those techniques that allow mapping deformations of the earth's surface in a very high resolution (i.e., millimeter-scale) [7] - [9]. These methods ideally complement terrestrial surveying.

To demonstrate the capabilities of InSAR technology, displacement maps of the Kurpsai Dam crest in Kyrgyzstan were created by [7] using TerraSAR-X radar data. They obtained a displacement of about -50 mm/year at the Kurpsai dam crest by measuring Line-Of-Sight (LOS) velocity on the selected point. Multi-sensor cumulative deformation map of Mosul Dam in Iraq generated from space-based SAR measurements from the COSMO-SkyMed constellation and Envisat satellites was studied by [10]. Their study results showed that the Envisat data measurements from 2004 to 2010 indicated a maximum LOS velocity of -11.5 mm/year and -12.47 mm/year vertical displacement, and the COSMO-SkyMed data analysis demonstrated a maximum LOS rate of -12.1 mm/year and -14.9 mm/year of vertical displacement at the Mosul dam during 2012–2015. With compared the profiles from 2004-2010 and 2014-2015, they found that there has also been a 300-millimeter shift to the east of the peak subsidence toward the dam's main spillway.

The current study has investigated the potential of the European Space Agency (ESA) Sentinel-1 data by applying the InSAR method to observe the deformation of the upstream face of the May embankment Dam. The InSAR technique has the potential to measure millimeter-scale changes in deformation over time spans. Therefore, InSAR applications for monitoring natural hazards, as well as structural engineering, monitoring of subsidence, and structural stability such as dam structure monitoring provide efficient information prior to the terrestrial measurements and help to make rapid decisions and prevent disasters.

The remainder of this paper is structured as follows: Section II presents the study area, data sources including a discussion of the nature and characteristics of the data, and the processing platform used in this study. This is followed by a description of the methodology used for processing the data. In Section III, we provide the results of our analysis and the key findings of the study. Finally, Section IV presents a discussion of the implications of the results and offers concluding remarks.

II. Materials and Methods

A. Study area

May Embankment Dam is located in the Çumra District of Konya province, at 37°31'19.27"N 32°32'23.71"E. It was built between 1957 and 1960 to provide irrigation services to an agricultural area of 1800 hectares as well as for flood protection purposes. The dam body is earth-filled, with a volume of 273,000 cubic meters and a height of 19.10 meters from the river bed. At normal water level, the volume and area of the lake are 40.10 cubic hectometers and 7.75 square kilometre, respectively. Figure 1 shows the location maps of the study area in Turkey and Konya province as well as the extension of the water behind the May Dam in the satellite image.

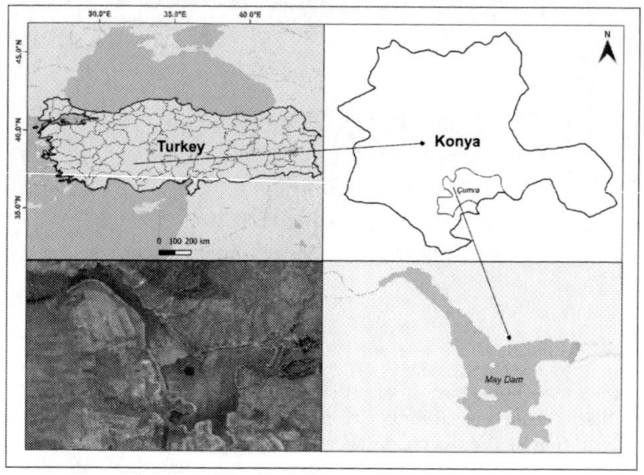

Figure 1. Shows the location maps of study area.

B. Sentinel-1 SAR data specification and processing

Sentinel-1 SAR data in ascending orbit pass, and in VV polarization were collected for the study area. For interferometric SAR and multitemporal analysis, a total of 8 SLC images in the Interferometric Width (IW) swath acquisition mode with a relative orbit (track number) of 160 were acquired from 2015 to 2022. Characteristics of the Sentinel-1 dataset used for the study and interferograms' information are given in Table 1.

TABLE I
Acquisition dates and characteristics of Sentinel-1 dataset and interferograms formation

Image No.	Abs.[a] Orbit-Frame	Acquition Date	Temporal Bas. (d)[b]	Perp.[c] Baseline (m)	Interferogram Formation
1	8282-119	23-Oct-2015			
2	13707-118	29-Oct-2016	372	25	2+1
3	18957-117	24-Oct-2017	360	-84	2+3
4	24382-117	31-Oct-2018	372	-18	3+4
5	29632-117	26-Oct-2019	360	177	4+5
6	35057-117	1-Nov-2020	372	-63	5+6
7	40307-117	27-Oct-2021	360	-110	6+7
8	45732-117	3-Nov-2022	372	107	7+8

[a]Absolute Orbit, [b]Temporal Baseline (days), [c]Perpendicular Baseline (meters)

The interferometric SAR processing is performed using the open-source tools of Sentinel Application Platform (SNAP) software [11]. The processing steps for obtaining the Sentinel-1 interferometric SAR maps are shown in Figure 2. The consecutive images in time (almost a one-year interval between primary and secondary image acquisition) were used for interferometric analysis. Figure 3 shows the baseline configuration related to the relative orbital number 160 in the ascending pass direction for the VV polarization. In Figure 3, the line represents the baselines between the consecutive images in time used for interferometric analysis and nodes corresponding to SLC images with acquisition date.

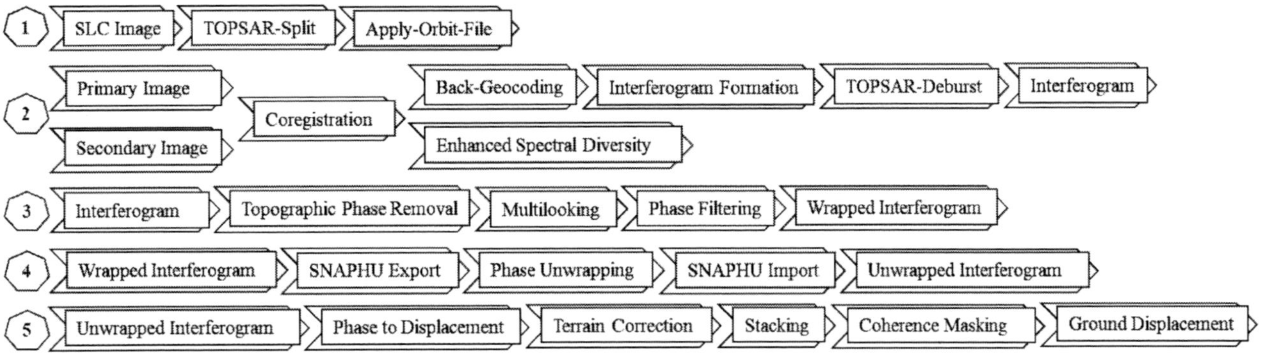

Figure 2. The pre-processing workflow schema for interferometric SAR processing of Sentinel-1 SLC SAR images.

In this study, to minimize the impact of atmospheric phase disturbances (atmospheric noise) and unwrapping errors, a series of images belonging to the same relative orbit (160) and polarization at a temporal interval of an average of 365 days were analyzed to obtain a more reliable surface displacement. All steps for the interferometric SAR processing were repeated for each consecutive image pair (i.e., 1+2, 2+3, 3+4, and so forth), resulting in a total of 7 geocoded displacement products between 2015 and 2022 (as shown in the last column of Table 1). The individual products are stacked in chronological order, then the average displacement is calculated using band math expression by adding the rasters for all dates and dividing the result by 7 (the number of interferograms). The same procedure was applied to obtain the average coherence maps to mask out parts of the image with low coherence to prevent misinterpretation of patterns caused by phase decorrelation. To mask areas of low coherence values from the image, all pixels with coherence values below 0.6 are eliminated and made transparent to better interpret the displacement map. However, before producing the stacked image, the geometric terrain correction algorithm using a Digital Elevation Model (DEM) is applied to correct the SAR geometric distortions and produce a map-projected product.

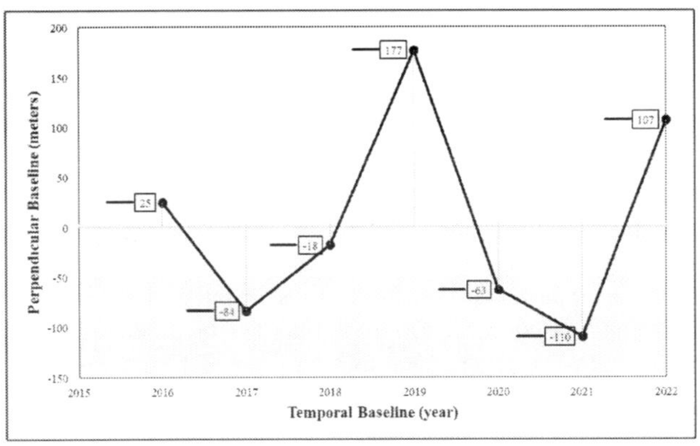

Figure 3. Baseline configuration used for SAR data processing. The horizontal axis is the acquisition time, and the vertical axis is the perpendicular baseline (distance between the two passes relative to the baseline).

C. InSAR Method

The InSAR technique is a method that reveals the topographic height, changes on the earth's surface or any object on the ground, and their spatial movements. With the InSAR method, the movement (velocity rates and vertical deformation) and the changes of the targets are obtained in sub-centimeter sensitivity in the radar viewpoint (i.e., LOS) direction and based on the phase differences (time delay between transmitted and received) of the radar signals from the SAR image pair that acquired at different time intervals from the same scene. In this method, after two radar images are co-registered and precisely overlapped to the considered resolution cell, the phase value of each pixel in the primary image (reference or master) is subtracted from the phase value of the corresponding pixel in the secondary image (slave). This creates a new image and is called an interferogram. In another word, the interferogram is generated by cross-multiplying the reference image with the complex conjugate of the secondary image. Accordingly, the computed interferogram contains phase variations $\Delta\phi_{int}$ from several influencing factors, the most important being the flat-Earth phase $\Delta\phi_{flat}$ (the curvature of the Earth), the terrain phase $\Delta\phi_{DEM}$ (topographic surface of the earth), atmospheric conditions $\Delta\phi_{atm}$ (humidity, temperature, and pressure change between the two acquisitions), and other noise $\Delta\phi_{noise}$ (variation of scatterers, different viewing angles and volumetric scatter), and lastly the final surface deformation $\Delta\phi_{disp}$ that occurs between two acquisitions "(1)" [12].

$$\Delta\phi_{int} = \Delta\phi_{DEM}+\Delta\phi_{flat}+\Delta\phi_{disp}+\Delta\phi_{atm}+\Delta\phi_{noise} \quad (1)$$

In addition to the interferometric phase, the coherence between the primary and secondary images is evaluated as an indicator of the quality of the phase information. Mainly, coherence shows whether the images have a strong similarity and thus are suitable for the interferometric processing uses. The Interferogram product is displayed in an interference pattern of color bands (as the rainbow pattern) called "fringes" resulting from the phase differences of two images and range from $-\pi$ to π. A fringe demonstrates a full 2π cycle and appears in an interferogram as cycles of arbitrary colors, with each cycle representing half the sensor's wavelength. Relative ground movement between two points is later derived by counting the fringes and multiplying by half of the wavelength, thereby, the closer the fringes are together, the greater the strain on the ground [13]. To be able to relate the interferometric phase that has ambiguous and known within the ranges of 2π to the terrain height, the phase must first be unwrapped. Phase unwrapping resolves the interferometric phase ambiguity with the change of 2π after interferogram flattening by integrating phase differences between neighboring pixels and removing any integer number of altitudes of ambiguity (equivalent to an integer number of 2π phase cycles). The phase variation between two points on the flattened interferogram provides a measurement of the actual altitude variation. Accordingly, the unwrapped product should be interpreted as a relative height/displacement between the pixels of two images.

To minimize the effects of atmospheric phase disturbances and unwrapping errors, SAR data from 2015 to 2022 were processed for each consecutive image pair. Then, the actual displacement pattern of the dam is calculated by averaging the displacements over several dates. In addition, using images from the dry season is a strategy that can be effective in minimizing the effects of atmospheric noise and residual phase changes. However, different methods, services and toolboxes are available for atmospheric noise correction, such as the Small Perpendicular Baseline (SBAS) [14] method, weather models (e.g., ECMWF ERA-I, WRF, NARR, etc.), spectrometer data (MERIS and MODIS), General Atmospheric Calibration Online Services InSAR (GACOS) [15], Toolbox for Reducing Atmospheric InSAR noise (TRAIN) [16], or a combination of different sources.

III. Results

The displacement map for the 7-year observation period (October 2015 to November 2022) of the May Embankment Dam shows zones of LOS motion on the dam's surface and its near area. Figure 4 shows the average deformation produced by the remote sensing InSAR method at the May Dam. The

Figure 4. The average deformation map of the May Dam measured by the InSAR method between 2015 and 2022 (Image source: Google Earth).

Sentinel-1 measurements from 2015 to 2022 indicate that the displacement at a linear rate of -9.5 mm/year (the negative rate indicates LOS increase or subsidence) on the upper part dam. However, time series analysis of the Sentinel-1 dataset shows a cumulative displacement rate of -22.16 cm for the period 2015–2022 (Figure 5). As it can be inferred from Figure 5, during the period covered by the Sentinel-1 dataset (2015-2018), the average deformation pattern was lower at about -2.1 mm/year. The reason for this is that between these years, the amount of precipitation in the area and catchment of the reservoir was low, and as a result, the amount of water behind the dam was either very little or even non-existent. The displacement rate started to increase between 2018 and 2020 due to the increased amount of precipitation during these years. The displacement rate for these years showed high increases and was obtained at -13.37 mm/year. Similarly, the displacement rate remained almost constant during 2020 and 2021 and started to increase again between 2021 and 2022 due to decreasing and increasing precipitation rates in the dam reservoir catchment. Figure 6 shows the time stamp of changes in the May embankment reservoir in the period 2011-2022, obtained from the Sentinel-2 dataset.

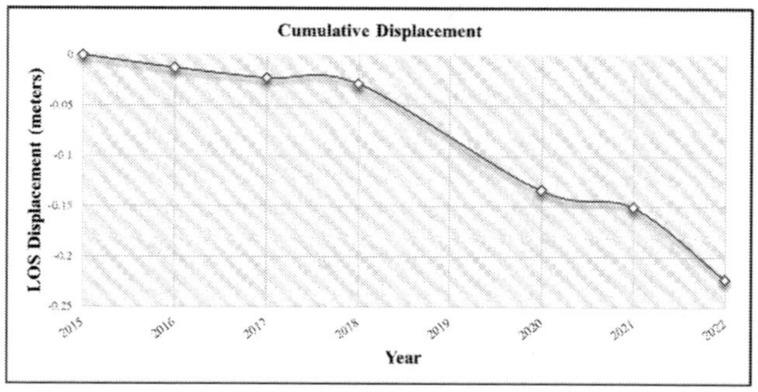

Figure 5. Time-series analysis, and the cumulative displacement on the dam's surface at the selected point.

In addition to the displacement of the upstream of the dam and its surrounding areas, in August 2002, three sinkholes were formed in the May dam reservoir that caused to loss of the dam's water in a short period of time. Figure 7 shows the location of the sinkholes in the May Dam reservoir. The sinkholes, which follow each other in a linearity, have 30-40m intervals from large to small. The diameters of the sinkholes are 5, 10-15, and 50 meters for the smallest to the large ones, respectively. Due to the increased hydraulic load caused by the rising water level in the dam lake, the ground pressure can increase and therefore the displacement process can be accelerated. Hence, apart from the potential displacement risks of the earth-fill structure of the dam, the deformation, and subsidence of the dam pond is also the subject. In 1960, the May Dam started to impounding, however, when the water height reached 6.7 m, the water of the dam suddenly dried up. The investigations revealed that 33 sinkholes had been formed at the dam reservoir. Once blocking the sinkholes failed, the dam has left self-abandoned [16].

Over the past two decades, ground depressions and sink-

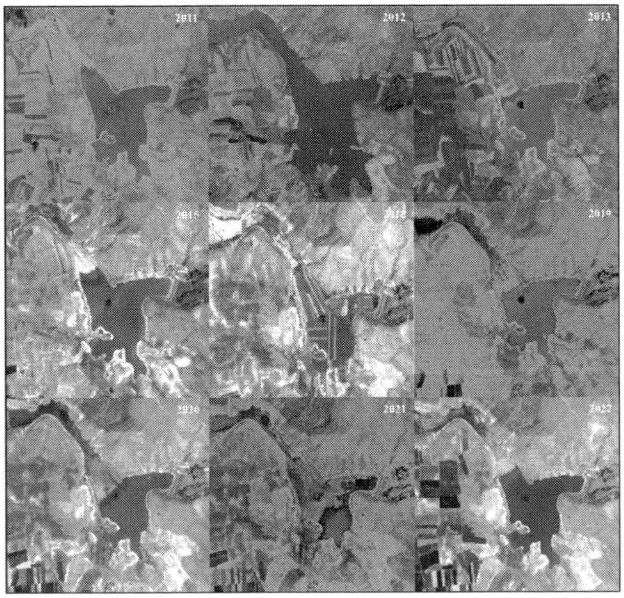

Figure 6. May Embankment Dam's reservoir variation during the time period of 2011-2022 (Imagery sources: Google Earth and Copernicus Sentinel data processed by ESA).

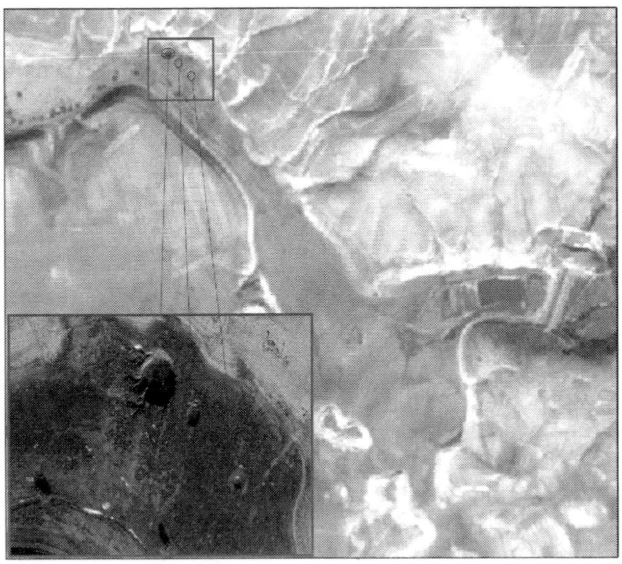

Figure 7. Images of the sinkholes formed in the May Dam (Imagery sources: Google Earth and Copernicus Sentinel data processed by ESA).

holes have begun to appear in the Konya region of Turkey. The land-use type in this area is mainly used for agricultural activities, and a large part of the water used to irrigate the farmland comes from groundwater. Hence, the main reason for the formation of the sinkholes and depressions in the region is likely due to the over-extraction of underground water.

This study has revealed the displacement rates of the structure of the earth-fill May Dam using the InSAR method. The results of this study can be used to warn the relevant authorities to make appropriate decisions and necessary measures to strengthen the dam to prevent potential risks. Due to inattention to the changes and displacements that occur in and around the dam structure, damage and failure of the dam can cause loss of life and property downstream of the dam.

This paper has presented an investigation into the deformation of the May Dam utilizing InSAR methodology. While the findings provide important insights into the current state of the dam and its potential risks, there remains a need for further research. As a next step, the correlation between the outcomes derived through InSAR methodology and ground-based measurements will be explored to validate the results. This approach is expected to enhance the reliability and accuracy of the findings, contributing to the broader understanding of the May Dam.

IV. Discussion and Conclusion

This study applied the potential use of the InSAR method to obtain the displacement rates of the May Embankment Dam structure. The displacement rate of the dam was estimated at a linear rate of -9.5 millimeters per year as a result of the InSAR post-processing analysis. However, the cumulative deformation rate of the 7-year processed SAR data (2015-2022) for the dam's structure is estimated at -22.16 cm. The rate of displacement increased between 2018 and 2020 due to the increased amount of precipitation and raising the water level of the dam reservoir during these years. The time series analysis of SAR data shows that the displacements occurring in the dam structure and its surrounding areas alert the authorities to take measures to prevent possible damage or failure of the dam and the occurrence of life and property losses at the dam's downstream. Although, different monitoring techniques can be employed to assess the stability and health of the dam and its safe operation, however, remote sensing methods provide highly accurate measurements with high temporal resolution. In conclusion, the InSAR technique along with high-resolution satellite SAR data can help monitor the displacements of the upper parts of large dams, it also provides the chance to extend the observed region to a large part of a structure instead of merely measuring a set of a few control points used by ground-based monitoring systems. Therefore, the InSAR technology can support the development of new and more efficient methods of monitoring and analyzing dams to complement land-based measurement approaches.

Acknowledgment

The authors would like to thank the colleagues and managers of the R&D and Quality and Management Systems Departments at Yüksel Proje International Co. for their contributions and efforts. The authors would also like to especially thank the management of Yüksel Proje (President and Vice President) and the Coordinator of the Water and Environment Department) for their support and encouragement of scientific and academic research.

REFERENCES

[1] Y. Kalkan, R. M. Alkan, and S. Bilgi, "Deformation Monitoring Studies at Atatürk Dam," FIG Congress, Sydney, Australia, pp. 11-16, 2010.

[2] Y. Dongang, W. Bowen, X. Bin, and Z. Zimeng, "Modified dam deformation monitoring model considering periodic component contained in residual sequence," Struct. Control Health Monit., 10.1002/stc.2633, 2020.

[3] Z. Yin and G.Gao, "Application of fractal theory to dam deformation forecast," Computer Modelling and New Technologies, 18(3), pp. 49-51, 2014.

[4] L. Tu, T. Bao, and Y. Li. "ARIMA Dam Early Warning Model Based on Fractal Interpolation," J. China Three Gorges Univ. Nat. Sci, 37, pp. 29–32, 2015.

[5] L. de, et al., "Monitoring of vertical deformations by means high-precision geodetic levelling. Test case: The Arenoso dam (South of Spain)," Journal of Applied Geodesy 11, no. 1, 11, pp. 31–41, 2017.

[6] A. Capra, M. Scaioni, and A. Wieser, "terrestrial remote sensing for areal deformation monitoring," Applied Geomatics 7, no. 2, pp. 61-63, 2015.

[7] O. Lang, W. Diana and A. Jan, "Satellite based long-term deformation monitoring on dams and its surroundings," In 5 th Conference on Smart Monitoring, Assment and Rehabilitation of Civil Structures, Pots dam, pp. 27-29. 2019.

[8] M. A. Ruiz-Armenteros, et al., "Deformation monitoring of dam infrastructures via spaceborne MT-InSAR. The case of La Viñuela (Málaga, southern Spain)," Procedia Computer Science 138, pp. 346-353, 2018.

[9] P. Riccardi, G. Tessari, D. Lecci, M. Floris, and P. Pasquali, "Use of Sentinel-1 SAR data to monitor Mosul dam vulnerability." In EGU General Assembly Conference Abstracts, p. 13098. 2017.

[10] P. Milillo, et al., "The ongoing destabilization of the Mosul dam as observed by synthetic aperture radar interferometry," IEEE International Geoscience and Remote Sensing Symposium (IGARSS), pp. 6279-6282, 2017.

[11] European Space Agency (ESA), Sentinel Application Platform (SNAP) v9.0.4, Available from: http://step.esa.int/ 2023.03.25.

[12] F. Rocca, C. Prati, A. G. Monti, and A. Ferretti, "InSAR Principles: Guidelines for SAR Interferometry Processing and Interpretation (ESA TM-19)," European Space Agency (ESA) Publications, February 2007.

[13] P. Berardino, F. Gianfranco, L. Riccardo, and S. Eugenio, "A new algorithm for surface deformation monitoring based on small baseline differential SAR interferograms," IEEE Transactions on geoscience and remote sensing 40, no. 11, pp. 2375-2383, 2002.

[14] Y. Chen, Z. Li, N. Penna, and P. Crippa, "Generic atmospheric correction online service for InSAR (GACOS)," In EGU General Assembly Conference Abstracts, p. 11007, 2018.

[15] D.P.S. Bekaert, R.J. Walters, T.J. Wright, A.J. Hooper, and D.J. Parker, "Statistical comparison of InSAR tropospheric correction techniques, Remote Sensing of Environment," doi: 10.1016/j.rse.2015.08.035, 2015.

[16] A. Doğu, "Falling of the three meteors into the May Dam," Available from: https://www.bildiris.com/ansiklopedi/May-Baraji/ 2023.03.25.

Geo-processing Approaches for Urban Water Supply and Drainage Systems' Data Rehabilitation

Cagri Cimen
Department of Water and Environment
Yüksel Proje Inc.
Ankara, 06610, Turkey
e-mail: ccimen@yukselproje.com.tr

Suleyman Canberk Tuskan
Department of Water and Environment
Yüksel Proje Inc.
Istanbul, 34485, Turkey
e-mail: sctuskan@yukselproje.com.tr

Anil Olgac
Department of Water and Environment
Yüksel Proje Inc.
Ankara, 06610, Turkey
e-mail: aolgac@yukselproje.com.tr

Abstract—**This study highlights the critical importance of accurate and comprehensive physical data in Integrated Urban Water Management (IUWM) and sustainable planning of water resources. Inaccurate or insufficient physical data can directly impact the reliability and sensitivity of planning, making it crucial for administrations to maintain accurate, thorough, and up-to-date data. To achieve reliable simulation results from hydrodynamic models, it is essential to define actual conditions in the model using correct and complete physical datasets. The study proposes a methodology for handling physical datasets of infrastructure systems in a GIS environment to obtain reliable simulation results. Digital and non-digital data from existing infrastructure systems are obtained and transferred to the GIS environment, with subroutines applied as needed to ensure accuracy and completeness. The physical data is then combined into a single database, correcting deficiencies and ensuring upstream-downstream continuity in the hydraulic model structure. With the integration of all physical data into the GIS environment, analysis and queries of geometric and attribute information are carried out for the detection and correction of deficiencies and errors of the system elements. This approach resulted in high-precision planning, detecting and correcting deficiencies and errors of the system elements, and ensuring the upstream-downstream continuity relationship in the hydraulic model structure. This study emphasizes the significance of accurate physical data in the sustainable management of urban infrastructure systems, and the proposed method can be utilized for any level of planning. Accurate and up-to-date physical data is essential for the sustainable management of urban infrastructure systems, and the use of the GIS environment and hydrodynamic models can lead to high-precision planning and reliable simulation results.**

Keywords—Water and drainage systems; hydrodynamic modeling; GIS analysis; data accuracy; sustainable water management.

I. Introduction

Water is a critical resource for human survival, and access to safe and clean water is essential for sustaining public health, environmental integrity, and economic prosperity. Efficiency and sustainability are primary goals of the services obtained by city municipalities and water companies in modern urban areas [1]. Integrated Urban Water Management (IUWM) methods, such as Sustainable Urban Drainage Systems (SUDS), Low Impact Development (LID), and Best Management Practices (BMP) have become increasingly widespread in recent years to achieve the Water-Sensitive City (WSC) goal. These methods aim to minimize the environmental impact of urbanization and improve the resilience of urban water systems to climate change and other stressors. Achieving a WSC requires more than just the implementation of these techniques. The multi-dimensional analysis capabilities of hydrodynamic modeling software must also be used as much as possible in the management and design of water infrastructure.

Model output simulation is a widely used method in modern cities, which can show problematic spots and risks in urban networks with high precision. However, accurate identification of problematic spots and potential risks in urban networks depends on precise, accurate, and up-to-date technical attributes of the entire urban network, which are typically stored in GIS databases [2]. Low input data quality can significantly affect the accuracy of simulation results. Data deficiencies are often encountered in GIS data for a variety of reasons, including converting data from Computer-Aided Design (CAD) or other formats into GIS. In particular, in metropolises, where a significant amount of data is being continuously entered into the system, ensuring data quality that can be used in hydrodynamic models to achieve accurate results is crucial [3]. Without high-quality input data, hydrodynamic models cannot be run without preprocessing, which can lead to deficiencies and errors in the simulation results [4] [5]. To address this issue, the ISO 19114 standard provides guidance for assessing and rehabilitating GIS data quality. The standard for data rehabilitation outlines a five-step process for improving data quality. The first step in the ISO 19114 standard is to assess the quality of the GIS data. This includes identifying any missing or faulty geometries, as well as missing data in the attribute table. Once the data has been assessed, the needs for data quality can be determined. This involves identifying the types of data that need to be improved, as well as the level of accuracy required. The next step is to develop a study methodology for improving the data quality. This involves identifying the appropriate tools and techniques that will be used to improve the data, as well as the resources that will be required to carry out the study. The study methodology will need to be tailored to the specific needs of the project, taking into account the type of data being collected, the level of accuracy required, and the available resources. Once the study methodology has been developed, the study results can be specified. This involves identifying the specific improvements that will be made to the GIS data, as well as the expected outcomes of the study. The study results should be communicated clearly to all stakeholders involved in the project, including water utility companies, city municipalities, and software developers. The final step in the ISO 19114 standard is to check the conformance of the data. This involves verifying that the GIS data has been improved to the required level of accuracy, and that it meets all relevant standards and guidelines. Checking the conformance of the data is essential to ensure that the hydrodynamic models run accurately and that WSC goals are achieved.

Hence, the ISO 19114 standard provides a valuable framework for the rehabilitation of GIS data in the context of IUWM. Following the standard can help to ensure that the technical attributes of the whole city network are kept up-to-date, accurate, and precise. By using engineering approaches to correct deficiencies in GIS data, and by verifying missing and certain system information with field studies, real-condition model results can be achieved. In addition, the successful operation of hydrodynamic models requires precise and complete technical attributes and upstream-downstream relationships of the network. This may require time and financial resources to complete, along with field studies to verify missing and incomplete system information. Nonetheless, correcting and rehabilitating the GIS data through engineering approaches in the GIS environment with site investigation support can lead to the accurate simulation results necessary for modeling a WSC.

The purpose of this paper is to highlight the importance of GIS data quality in hydrodynamic modeling and to discuss engineering approaches to improve data quality in a GIS environment. The proposed method aims to improve the accuracy and reliability of hydrodynamic models and provide more reliable decision support for water infrastructure management and design.

The subsequent sections of this paper are structured as follows: Section II outlines the problems encountered and the methods employed to address them, including a presentation of the methodology. Section III provides an in-depth explanation of the problem and the approach utilized. The methodology is also presented in Section II. Finally, Section IV presents the conclusion of the paper, summarizing the benefits and contributions of this work.

II. METHODOLOGY

In the context of hydrodynamic modeling, accurate and complete GIS data is critical for achieving reliable results. This is particularly important for water infrastructure management and design, where incorrect or incomplete data can lead to costly errors and inefficiencies. One common challenge in hydrodynamic modeling is ensuring the continuity of manholes and pipelines in the software. The hydrodynamic modeling software possesses GIS software capabilities, allowing it to represent both wastewater and stormwater lines utilizing point geometry for manholes and line geometry for pipelines. However, in actuality, these manholes and lines are interconnected, and it's necessary to have no gaps between points and lines in the software. Furthermore, since the wastewater and stormwater lines rely on the gravitational movement of water, thus, to address this issue, it is essential to ensure manhole-line-manhole continuity in both plan and profile from the upstream point to the final downstream point. Data deficiency and gaps should be identified and completed according to this principle. To achieve this, classifications can be made to identify problems, such as non-connected lines and manholes with short distances that are not visible to the eye and faulty or missing lines (Figure 1).

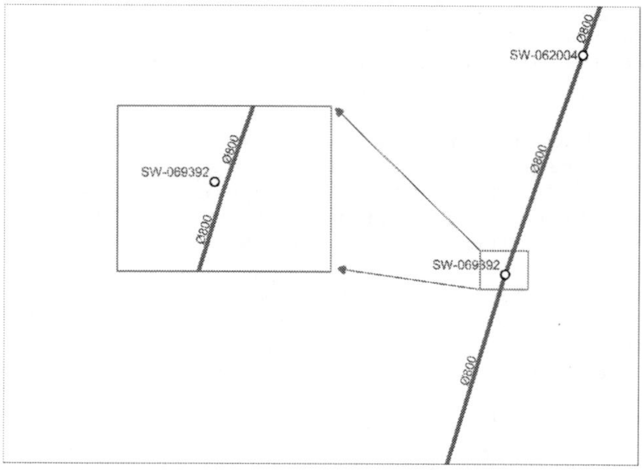

Figure 1. Sample of non-connected point (i.e. manhole) with the lines (i.e. pipes) error.

Non-connected manholes and lines can be caused by issues, such as data loss during the transfer of data from field acquisitions, sensitivity mismatch, and other factors that can disrupt data integrity. Faulty or missing lines may occur due to post-measurement data transfer problems, database errors, and other issues. To overcome these challenges, engineering approaches can be employed to improve GIS data quality in the hydrodynamic modeling process. By addressing these challenges and improving GIS data quality, more reliable hydrodynamic models can be developed, providing better decision support for water infrastructure management and design.

In the preliminary evaluation of the geometric and attribute information in the point (manhole) and line (pipe) layers for the use of the data in hydraulic analysis, in this study, two main problems were identified:

- discontinuity/disconnection (that is not visible) between the pipe and the manhole,
- broken and missing pipelines (on a visible scale).

The identified problems are classified and resolved according to the developed algorithm explained in detail below.

A. *GIS Data Quality Improvement Approaches*

The obtained data have been subjected to two types of geoprocessing analysis and evaluation in GIS applications.

a) Spatial Analysis

- Intersect analysis
- Near analysis
- Select by location
- Select by attribute
- Buffer

b) Topological Analysis

- Point-Line Relation Topological
- Must be Covered by Endpoints
- Must be Covered by Line (Snap, extend, trim)
- Point-Point Relation Topological

The spatial analysis was employed to conduct intersection analysis in order to obtain information about the manholes that were in contact with the lines. Proximity analysis was used to transfer information on manholes located within a certain distance to the line database. Various analyses, such as connected/not-connected and coverage, were carried out using the Select by Location tool. The identification of related lines and manholes was accomplished through the Select by Attribute tool, which involved a comparison of the attribute tables. The topological analysis of the data involved checking whether the line and manhole layers were complementary to each other. The analysis verified the presence of manholes at both ends of the line and the existence of a line touching the manhole. Through the use of geoprocessing algorithms, such as "Spatial Analysis" and "Topological Analysis" in the GIS environment, physical disconnections and numerous deficiencies of the same type, which could not be detected visually, were identified.

These approaches have resulted in improvements to both the point and line layers, correcting errors on a scale that may not be visible to the naked eye. The main errors and shortcomings have been categorized as follows:

Point (Manhole) Layer:

- Non-connected points with the line
- Missing points (manholes) at the start or end of the lines

Line (Pipe) Layer:

- Pipes that are a continuation of each other but whose ends do not touch each other
- Duplicate lines
- Missing data and errors that are visible at the line layer, such as the case of missing and disconnected collector lines or the case of missing discharge lines.

B. Disconnected Lines and Missing data (Visible errors)

In some cases, certain lines in the GIS database may not accurately represent their real-world counterparts. This can occur due to a variety of factors, including difficulties with fieldwork and challenges in properly adding geometry and attribute data for older lines. Additionally, issues may arise when attempting to convert CAD data of existing lines to the GIS environment. These conversion problems can result in missing data and errors in the GIS database, which can lead to inaccuracies in the hydrodynamic models. To address these issues, engineering approaches have been employed in a GIS environment with the oversight of local administration controls. Corrections were made to the GIS database to improve the accuracy and reliability of the input of these data into the hydrodynamic model through careful inspection and analysis to identify errors that were missed. This process has involved a combination of techniques, including manual updates to the GIS database and the use of geoprocessing algorithms to identify and correct issues.

III. Proposed Solutions

In the analysis results, several error types are commonly observed. To ensure proper control and tracking of analysis and queries performed on the GIS database, it is recommended to create a new column in the attribute table and record the corresponding operations. The different types of errors and their respective arrangement methods are discussed below:

1. Disconnection of points and lines: This error occurs when a line is missing in the upstream-downstream route (Figure 2). To address this issue, the missing lines are completed to establish the upstream-downstream relationship up to the discharge point. The completion process takes into account estimated information about the systems, topographical bases, and current maps obtained in one-on-one meetings with administrative units.

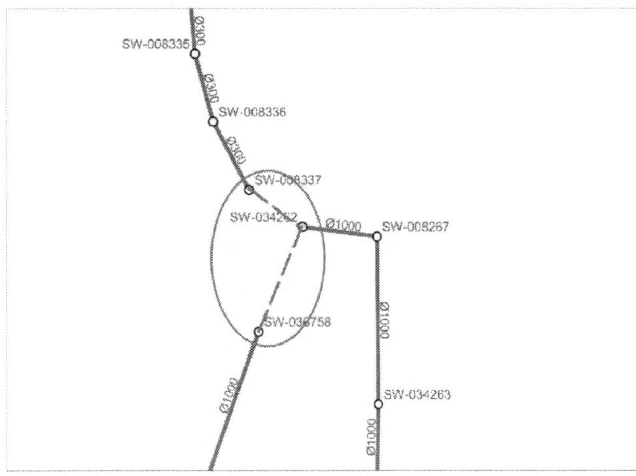

Figure 2. Sample error of the disconnection of points and lines (i.e., disconnection of manholes with lines).

2. Missing discharge lines: This error refers to lines that lack an outlet line extending to the final discharge point (Figure 3). To resolve this issue, the line is extended from the last manhole to the discharge point, vertically connecting it from the shortest distance. The diameter of the last pipe is considered, and if the discharge point is a stream, the bottom elevations of the stream are taken into account. If the discharge point is a lake or sea, the slope of the previous line is considered.

3. Indeterminate cross-section dimensions of lines: This error occurs when cross-section dimensions are undefined or empty in the attribute table. The deficiency is completed by assuming that it is the same size as the previous line.

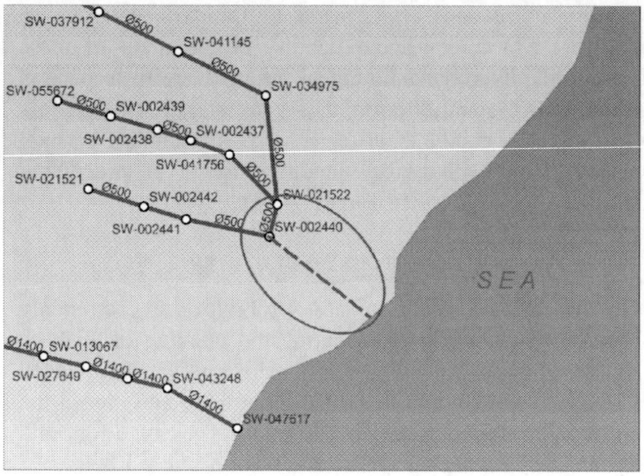

Figure 3. Sample error of the missing discharge line.

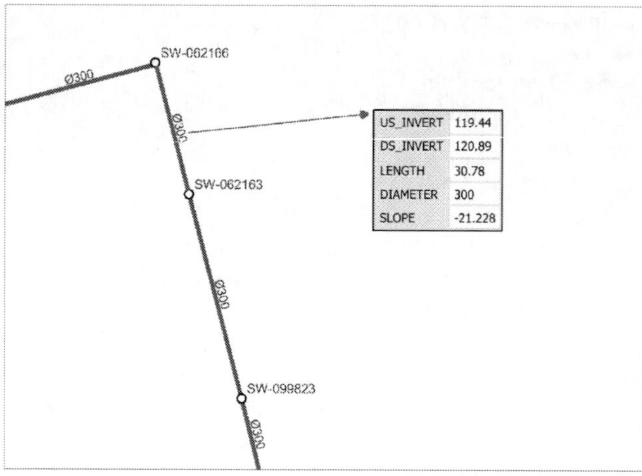

Figure 4. Sample error of the reverse slope of the pipeline.

4. Line elevations not defined: This error represents lines that are not determined, defined as 999.99, or as the same as manhole ground elevations. To address this error, the deficiency is completed by considering the flow levels of the previous/next pipe and manhole. If information is lacking for more than one manhole and/or line, the nearest manhole and/or lines with available information were used.

5. Undefined ground level in manhole: This error refers to manhole floor levels that are undefined or defined as 999.99. To complete the missing data, the Digital Elevation Model (DEM) is used.

6. Manhole ground level and base level being equal: This error occurs when the manhole floor level is incorrectly the same as the ground level. To resolve this issue, the ground level is first checked on the digital elevation model, and then the completion method #4 is applied.

7. Pipe ridge elevation > ground elevation: This error occurs when the pipe ridge elevation is above the ground. If the drain line is acceptable, no further action is required. Otherwise, the flow level and diameter compatibility at the beginning and end of the pipes is checked. If these conditions are not met, the ground level is checked, and the arrangement is made accordingly.

8. Two lines output from one manhole: It means that there are two lines downstream of the manhole. (When a collector divides into two or more branches, the flow calculations become more complex, and as a result, some modeling software may not be capable of solving such situations)

Arrangement method: If the capabilities of the modeling software used can solve this situation, it is used exactly, otherwise, it is reduced to a single line.

9. Reverse slope of the lines: The elevation of the stream on the upstream side of the line is lower than that on the downstream side (Figure 4).

Arrangement method: Correction for deficiencies no. 4, 6, and 7 is applied on reverse sloping lines, otherwise, it is accepted as is.

10. Zero-slope lines: These are the lines with equal stream elevations upstream and downstream (Figure 5).

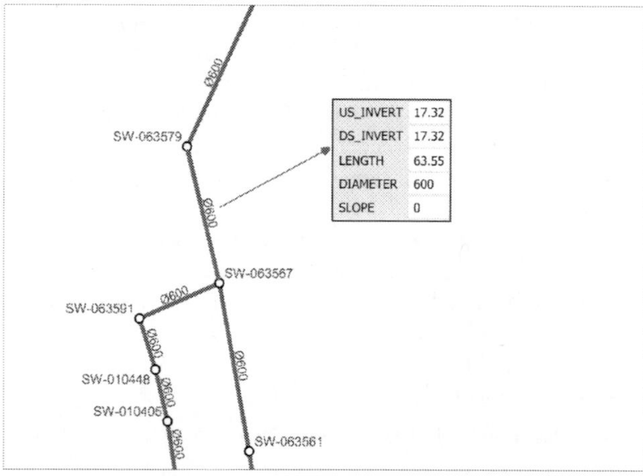

Figure 5. Sample error of the equal flow levels of the pipeline.

Arrangement method: Correction for deficiencies 4, 6, and 7, such as reverse slope lines is applied, otherwise, it is accepted as is.

11. Single lines: Usually refers to culverts and road crossings.

Arrangement method: Since single lines with large cross-sections are usually produced at road crossings, there are no upstream and downstream lines. Therefore, they are not considered as part of the collector system.

12. Multiple points at close range represented by a single manhole: Large cast-in-place manholes represented by multiple point objects with the same characteristics.

Arrangement method: If this manhole definition is made as 2 or 3 points, these points are arranged to be defined with a single point for the hydraulic model base.

13. Small section lines between large section lines: Small section lines connecting two large section lines (Figure 6).

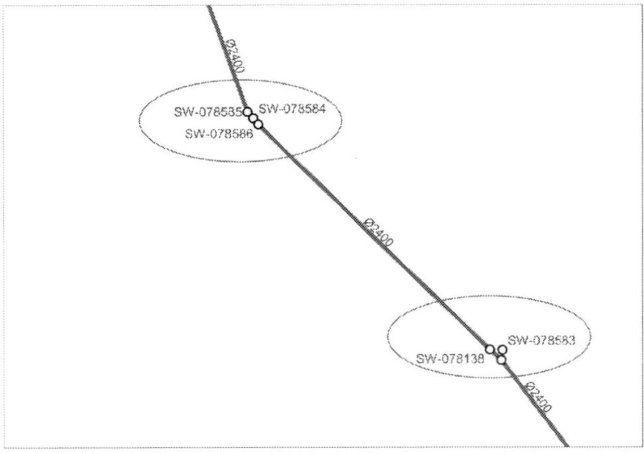

Figure 6. Sample error of the multi-point display at near-distance of cast manholes.

Arrangement method: The section size compatibility is checked by examining the entire line. In this context, if the difference between the dimensions is large, the diameter is considered large; otherwise, it is preserved.

14.Repeating or crossing lines on the same route: It is the situation where there are lines of different or the same size on the same axis or road (Figure 7).

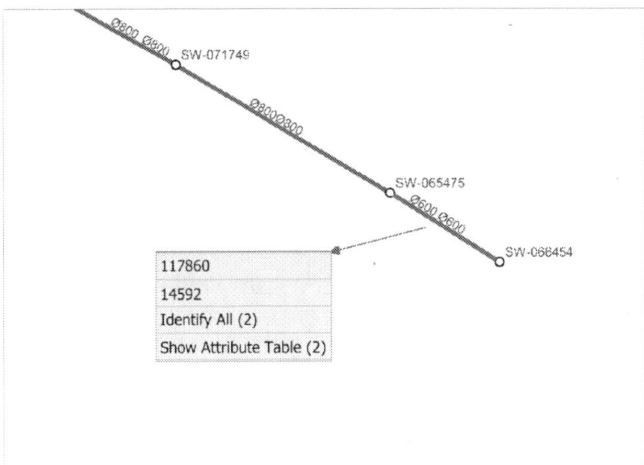

Figure 7. Sample error of repeating or crossing lines on the same route.

Arrangement method: If there are lines starting from the same upstream point, manufactured on the same axis, on different dates, and cut in different diameters and discharged to the same point, the larger diameter line or the newly constructed line is preserved.

To overcome the problems experienced during the transfer of planning from the CAD environment to the GIS environment, it may be advantageous to use middleware that we develop ourselves. The Autolisp platform, which is Autocad's macro development interface, stands out with its widespread awareness and capabilities.

Figure 8 shows the algorithm to convert an infrastructure plan in an Autocad file to an Excel sheet of manholes, along with their attribute information using the AutoLISP Visual Basic code.

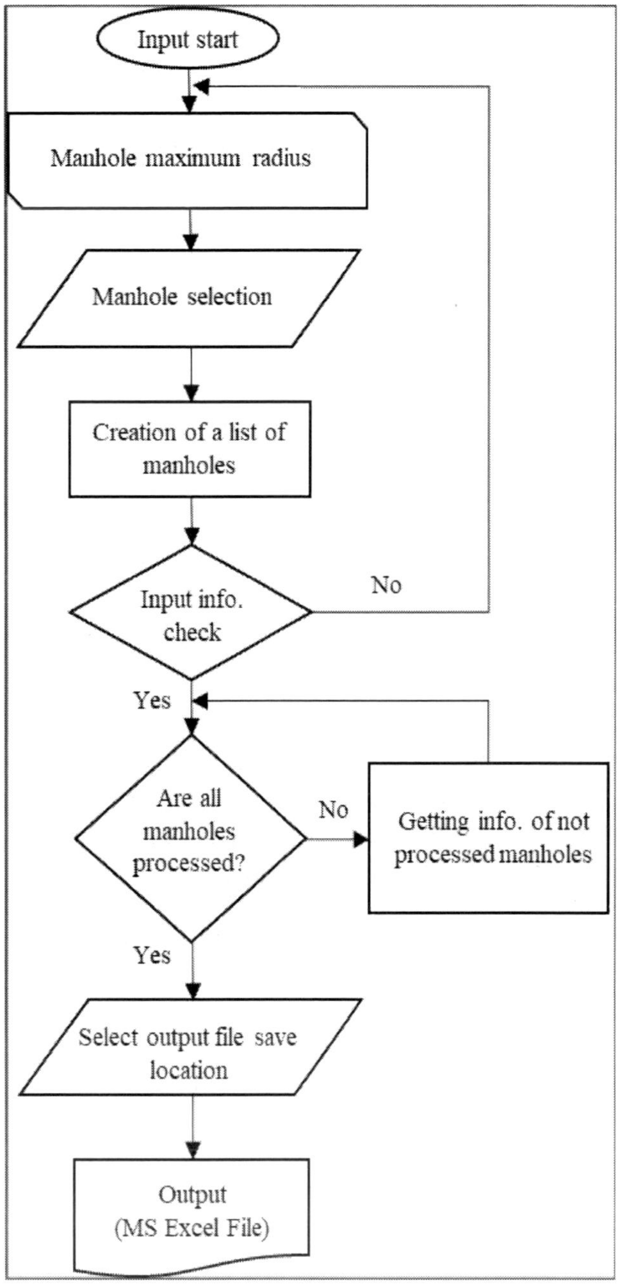

Figure 8. Flowchart to convert an Autocad file to an Excel sheet of manholes.

The algorithm presented here extracts essential information from manholes in a sewer network using the AutoLISP Visual Basic code. Specifically, the code captures the coordinates of the manholes from the center of the circle, as well as their respective manhole numbers, ground elevation, and flow elevation information from the texts written on the manhole.

The explanation of each step of the algorithm is provided below.

1- Start Lisp: The program starts by invoking Lisp.
2- Get User Input "Enter the maximum manhole radius": This step prompts the user to enter the maximum manhole radius that is required for the separation of manhole and label circles.
3- Get User Input "Select Manholes": This step prompts the user to select the manholes for which data needs to be exported to an Excel file.
4- Creates a list from manhole circles: This step creates a list of manhole circles that have been selected in step 3.
5- Suitability check for the Radius and Selection inputs: This step performs suitability checks for the maximum manhole radius and the selected manholes. If the input values are not suitable, the program returns to step 2. If the input values are suitable, the program proceeds to the next step.
6- Are all manholes processed? (Loop): This step checks whether all the manholes have been processed in a loop. If all manholes have been processed, the program proceeds to step 7. If not, the program extracts the X, Y coordinates, manhole number, ground level, and manhole invert level from the manhole label texts and proceeds to the next iteration of the loop.
7- Select a location for save data to an excel file: This step prompts the user to select a location to save the data to an excel file.
8- Write Manhole data list to Excel file: This step writes the list of manhole data to the Excel file that has been selected in step 7.

Figure 9 shows the sample of the extracted information and transferred to an Excel file. The the output file can be easily converted to a GIS shapefile format, allowing for efficient data analysis and further processing in the GIS environment.

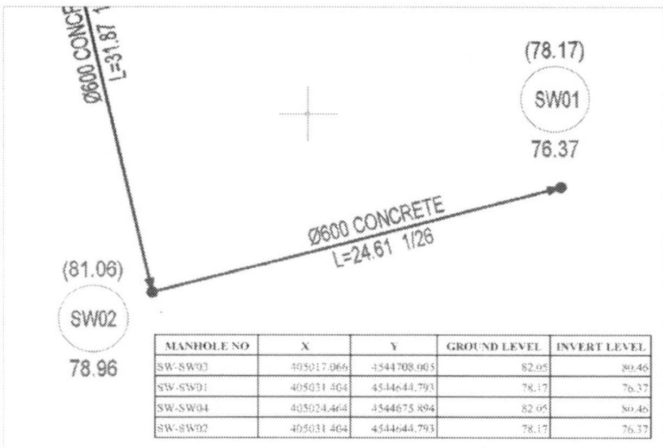

MANHOLE NO	X	Y	GROUND LEVEL	INVERT LEVEL
SW-SW03	405017.066	4544708.065	82.05	80.46
SW-SW01	405031.404	4544644.793	78.17	76.37
SW-SW04	405024.464	4544675.894	82.05	80.46
SW-SW02	405031.404	4544644.793	78.17	76.37

Figure 9. Example of transferring manholes CAD data to excel with Autocad Visual Basic lisp.

IV. Conclusion

The proper design and modeling of urban water supply and drainage systems are essential for urban planning and infrastructure development. Effective hydrodynamic modeling requires careful consideration of multiple factors, such as data collection, model setup, calibration, and validation. Accurate data collection is essential for developing a reliable model, and it involves the use of appropriate data quality improvements approaches, such as data cleaning, filtering, and normalization. The editing and arrangement of data help to ensure the quality of the model and avoid errors. The correction and validation of input data are also critical stages that enable the model to represent the real-world system accurately. This requires a thorough understanding of the hydraulic and geometric properties of the system components and the use of appropriate software tools to model and analyze the system's behavior. The presented guidelines and data quality improvement methods in this paper provide a useful framework that helps to ensure the accuracy and completeness of the data used in the modeling process. Furthermore, the use of middleware, such as Autocad Visual Basic code can streamline the data transfer process between different software environments and facilitate the integration of data into a GIS platform. GIS tools and algorithms provide a range of benefits, such as data visualization, spatial analysis, data editing, and the ability to integrate multiple data sources. This integration enables efficient visualization and analysis of the urban water supply and drainage system's behavior and performance, aiding decision-making processes related to system design and management.

Acknowledgment

The authors would like to thank the colleagues and managers of the R&D and Quality and Management Systems Departments at Yüksel Proje International Co. for their contributions and efforts. The authors would also like to especially thank the management of Yüksel Proje (President and Vice President) and the Coordinator of the Water and Environment Department) for their support and encouragement of scientific and academic research.

References

[1] S. Han, H. Hwang, S. Kim, G. S. Baek, and J. Park, "Sustainable water infrastructure asset management: a gap analysis of customer and service provider perspectives," Sustainability, 7.10, pp. 13334-13350, 2015.
[2] M. Badhrudeen, E. S. Boria, G. Fonteix, M. D. Siciliano, and S. Derible, "The C2G Framework to Convert Infrastructure Data from Computer-Aided Design (CAD) to Geographic Information Systems (GIS)," Informatics. Vol. 9. No. 2. MDPI, 2022.
[3] H. Baumann, N.H. Ravn, and A. Schaum, "Efficient Hydrodynamic Modelling of Urban Stormwater Systems for Real-Time Applications," Modelling 3.4, pp. 464-480, 2022.
[4] T. Richta, "Issues of GIS data management," Geoinformatics Fce Ctu 1, pp. 56-63, 2006.
[5] N. Mardani, et al., Sidle "Improving the accuracy of hydrodynamic model predictions using Lagrangian calibration," Water 12.2 p. 575, 2020.
[6] G. Droj, S. Suba, and A. Buda, "Modern techniques for evaluation of spatial data quality. revcad," Journal of Geodesy and Cadastre, 9, pp. 265-272, 2009.

Development of a Web-Based Geospatial Application for Efficient Spatial Data Management

Rouhollah Nasirzadehdizaji
Department of Water and Environment
Yüksel Proje Inc.
Ankara, 06610, Turkey
e-mail: rnasirzadeh@yukselproje.com.tr

Anil Olgac
Department of Water and Environment
Yüksel Proje Inc.
Ankara, 06610, Turkey
e-mail: aolgac@yukselproje.com.tr

***Abstract*—Today, digitization and dynamic representation of geospatial data generated on different Geographic Information System (GIS) platforms are mainly performed by location-based information applications called Web-based GIS (Web-GIS). Web-GIS technology encompasses multiple levels of technology for publishing maps on the web, ranging from simply publishing static web maps to dynamically serving maps, interactive custom maps, and more complex sites that support multiple computer platforms and operating systems. In this work, a Web-GIS application has been designed and developed by using different programming languages (and libraries) and implemented by providing interoperability between different open-source software, which form the basis of the Web-GIS application such as GIS software database/GIS data entry interface, and map server. A Web-GIS application that can be accessed from a computer or mobile device, allowing the dynamic presentation, analysis, and manipulation of spatial or non-spatial data for various purposes. Furthermore, it can be applied to quickly access data obtained from different disciplines within an institution and stored in databases from anywhere in the world if the user can connect to the Internet. Web-GIS applications can be used as an efficient decision support mechanism for decision makers and authorities for quick and accurate decisions because there is no need for data pre-processing, and all the spatial data stored in the system are post-processed and the data displayed in a user-friendly interface with the arranged styles. While Web-GIS applications are designed to be accessible and user-friendly, it should be recognized that some level of GIS knowledge may still be necessary to fully interpret and understand geospatial data. This work's significance lies in its demonstration of the integrated use of open-source software and interoperability between them to develop a Web-based GIS application that caters to the needs of decision-makers and authorities, promoting collaboration and data sharing within organizations and institutions.**

Keywords*—Dynamic and interactive maps; geographic information system; geospatial database; Web-GIS applications; web mapping.

I. Introduction

The geospatial data contains general (maps) and specific (attributes) information about a data (i.e., layer). The development of computer and Internet technology has also accelerated the speed of accessing graphical and geospatial data using Web-GIS applications. A Web-GIS system is a kind of internet-based GIS application that uses an online map and presents spatial data on the web. Several terms are commonly used in the context of Web-based GIS, including Desktop GIS, Distributed GIS, Internet GIS, and Web-GIS. These terms are all types of GIS services, however, they differ in their characteristics and functionalities. Desktop GIS is a powerful GIS software that is installed on a desktop computer, requiring specialized knowledge and training in GIS. Distributed GIS stores spatial data in distributed locations, enabling multiple users to access and work collaboratively on spatial data. Internet GIS uses the Internet to share spatial data and is commonly used for disseminating information related to environmental planning, disaster management, and public health. Web-GIS, on the other hand, is a user-friendly GIS system that is accessed through a web browser and used for accessing and analyzing spatial data related to environmental planning, urban development, and public infrastructure management.Therefore, Desktop GIS, Distributed GIS, Internet GIS, and Web-GIS are all types of GIS services with different characteristics and functionalities, designed for specific purposes and user groups.

The terms Web-GIS and web mapping are used synonymously, even if they do not mean the same. In fact, the boundary between web maps and Web-GIS is blurry. Web-GIS uses web maps and provides analytical capabilities to web mapping end users. A web map typically includes a web browser or other user agent capable of performing client-server interaction. Web maps are a presentation environment for Web-GIS, and are gradually gaining analytical capabilities. In other words, web mapping is the process of designing, implementing, and presenting maps provided by GIS on the World Wide Web (WWW) and its products [1] [2].

Web-GIS applications are developed for many different purposes as they serve and provide advantages in various fields. Tourism, environment, e-commerce, managing land use, and planning and designing transportation systems are the most widely used areas of Web-GIS applications. The Web-GIS application is considered to be the first system to meet the needs of tourism [3]. The easy-to-use Web-GIS applications attract environmental scientists and decision-makers to better understand the spatial, socioeconomic, and historical aspects of the hazards and the associated risks to society and to increase people's awareness of these risks [4]. E-commerce companies and the private sector benefit from the potential of Web-GIS to present and promote themselves. Even though 2D and 3D Web-GIS systems can be applied to different projects and fields, it provides more opportunities to increase the use of these applications [5] [6].

Recent studies have explored the potential of Web-GIS applications in a variety of fields. For example, [7] developed a Web-GIS approach for evaluating the effects of green infrastructure on urban thermal environment. The authors used a combination of remote sensing data and field measurements to assess the thermal performance of different types of green infrastructure in an urban setting. The resulting Web-GIS platform enabled users to visualize the spatial distribution of temperature and vegetation cover, and to compare the thermal performance of different green infrastructure types. The authors concluded that the Web-GIS approach could provide valuable insights into the effectiveness of green infrastructure in mitigating the urban heat island effect.

In another study, [8] developed a Web-GIS platform for monitoring and predicting air pollution in urban areas. The authors used a combination of satellite data and ground-based measurements to create a real-time air quality monitoring system. The resulting Web-GIS platform provided users with up-to-date information on air quality at the neighborhood level, and enabled them to visualize the spatial distribution of air pollutants. The authors concluded that the Web-GIS platform could be a valuable tool for urban planners and policymakers in developing strategies to mitigate air pollution and improve public health.

In the most recent work, [9] proposed a web-GIS based approach for flood risk assessment and management in urban areas. The authors developed a web-GIS platform that integrates various data sources such as topographic data, hydrological data, and land use data, to identify flood-prone areas in urban regions. The platform also includes a decision support system for flood risk management, which allows users to explore different scenarios and evaluate the effectiveness of various flood management strategies.

The aim of this work was to develop a Web-GIS application for geospatial data management, analysis, and presentation. Specifically, the methodology is demonstrated through the development of a Web-GIS application that can provide a platform to better manage, analyze, and present large datasets in the web environment for efficient use of processed data. While previous studies have used Web-GIS applications for similar purposes, our methodology provides a systematic and efficient approach to creating such applications that can be applied to a wide range of environmental projects. This contribution is particularly valuable in the context of increasing interest in using geospatial technologies to manage and analyze and present environmental data in an efficient manner, making it a useful tool for decision support systems in a variety of environmental projects.

The paper is organized as follows: Section II describes the methodology for developing the Web-GIS application, including the system development process and the key features of the developed application. Section III provides a discussion of the potential benefits of using the developed Web-GIS application. Finally, Section IV concludes the paper and suggests avenues for future research.

II. METHODOLOGY

This section provides an overview of the methodology used in the development of the Web-GIS application. The following sub-sections describe the different phases of the development process in detail, including system design and implementation, and interface development. The software used and their interoperabilities are also discussed to provide a better understanding of the development process.

A. System Development

The global nature of Web-GIS is inherited from the widely supported Hypertext Transfer Protocol (HTTP). Almost all organizations open their firewalls on certain network ports to allow HTTP requests and responses to pass through their local networks, thus increasing accessibility. Web-GIS can be designed as a web-based mapping application with multi-platform features, easy to use for end users, and compatible with different web browsers, such as Google Chrome, Internet Explorer, and Firefox. Web-GIS based on HyperText Markup Language (HTML) clients typically supports different operating systems (e.g. Win, Linux, Mac OS, iOS) as web browsers largely adhere to HTML and JavaScript standards. In a web-GIS application, information exchange takes place between a server and a client, where the server is a GIS server and the client is a web browser, mobile application, and desktop application. The server has a unique "Uniform Resource Locator" (URL; used to show the location of a resource on the internet) so that clients can find this information on the web and perform various actions. Figure 1 shows the main components of the Web-GIS application system and the interaction between the components.

Within the scope of this study, different processes were carried out for the realization of the Web-GIS system. Depending on the purpose of the Web-GIS project, where a server system is provided, the server needs to provide a specific URL on the web so that clients can easily access it. Clients rely on the HTTP specification to send requests to servers. The server performs the requested GIS operation and sends a response to the client via HTTP. The format of response sent to the client can be in many formats such as HTML, binary image, JavaScript Object Notation (JSON), and Extensible Markup Language (XML). The software and hardware infrastructure has been created for the project to be displayed on the web, including server setup, user account, and application access management, provision of security policies, interface design of applications and transactions, etc. In the design of the Web-GIS application, establishing a database for entering, processing, and keeping spatial data is carried out for the processes of importing the GIS data into the spatial database, maintaining the existing databases, and publishing the data on the web. For the establishment of a map server to display spatial data on the web, the map server needs to be set up to convert the selected spatial data (maps) into a set of display elements and make the created maps visible to the user on a suitable display device. Styles such as symbols, lines, and fill styles are added to spatial properties, annotations are created from alphanumeric

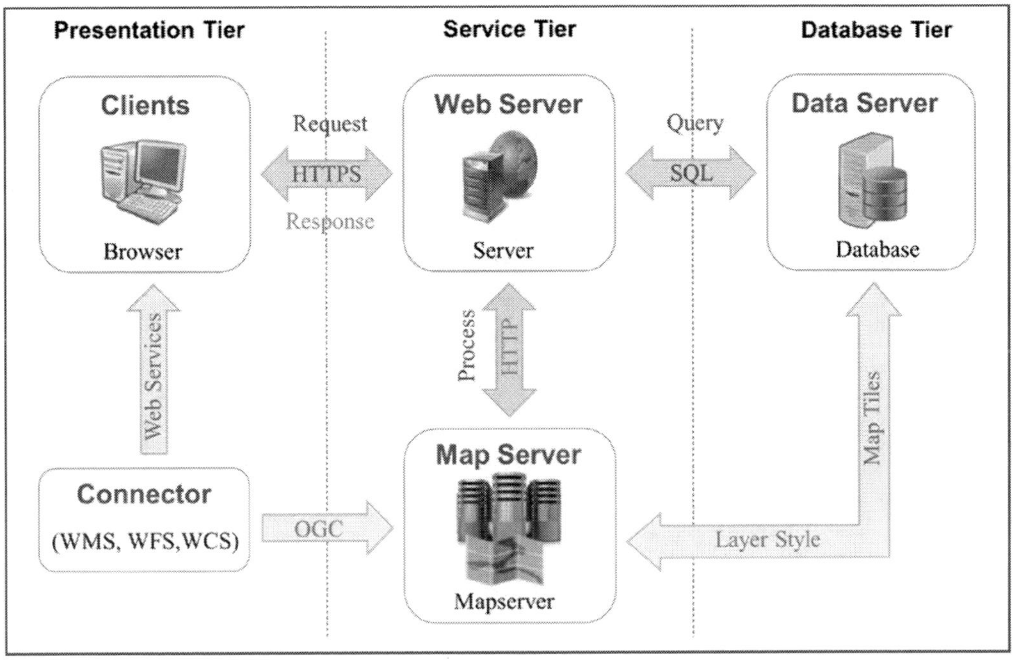

Figure 1. The structure and main components of the web-GIS application system.

properties, image elements are sorted in a specific order, and other graphics operations take place within the map server process. The features and extents obtained from the selection process are prepared from data acquisition and conversion into a format (raster or vector) suitable for processing into a database. Generated data and maps are prepared in GIF files or postscript files. In this work, open source software that forms the basis of the Web-GIS system, such as JavaScript and QGIS, PostgreSQL/PostGIS, and GeoServer are conducted to develop the Web-GIS application by providing interoperability between GIS software, database/GIS data entry interface and map server, respectively. Features used as spatial data can also be in different formats, such as vector and raster layers (Shapefile and GeoTIFF) or text (XML) data types. GeoServer functions as the Web Feature Service (WFS) reference implementation of the Open Geospatial Consortium (OGC) standard, as well as using the Web Map Service (WMS), Web Coverage Service (WCS), and web render service features.

B. The Developed Web-GIS Application

The developed Web-GIS application in this work can be accessed through any web browser and can be accessed from the URL [10]. The application login screen and interface components are shown in Figure 2. The components are numbered and each component is thoroughly explained below.

The username/password entry screen (number 1) is the first part that appears after the user/customer opens the application via the web browser. The login part of a Web-GIS application is a critical component that plays a crucial role in ensuring the security and privacy of the data stored in the application. It allows users to authenticate themselves and gain access to the different features and functionalities of the application based on their level of authorization. The importance of the login part lies in its ability to categorize users on the back-end of the application in terms of storing existing, confidential, and undesirable data in the database. This categorization enables the application to store sensitive data and confidential information in a secure and protected environment, inaccessible to unauthorized users. Additionally, the login part helps in managing user access to the application, restricting or granting permissions based on the user's role or responsibilities, ensuring that users only access the information and data that is necessary for their tasks. Overall, the login part of a Web-GIS application is a critical feature that ensures the security, confidentiality, and integrity of the data stored in the application, while also enabling users to access the features and functionalities they need to perform their tasks.

After successfully logging into the Web-GIS application, the user can move on to the second component (number 2), where they select the data group they want to examine (i.e., the spatial query section). The spatial query component plays a critical role in enabling users to navigate through the vast amounts of data stored in the database and search for specific information. The spatial query help filter the data within a specific group or project, enabling the user to focus on the most relevant and important information. In another word, a user working on a project may want to filter the data to show only the data that meets specific criteria or conditions. The pre-defined conditional rules as the Structured Query Language (SQL) modules in the application enable the user to do the spatial

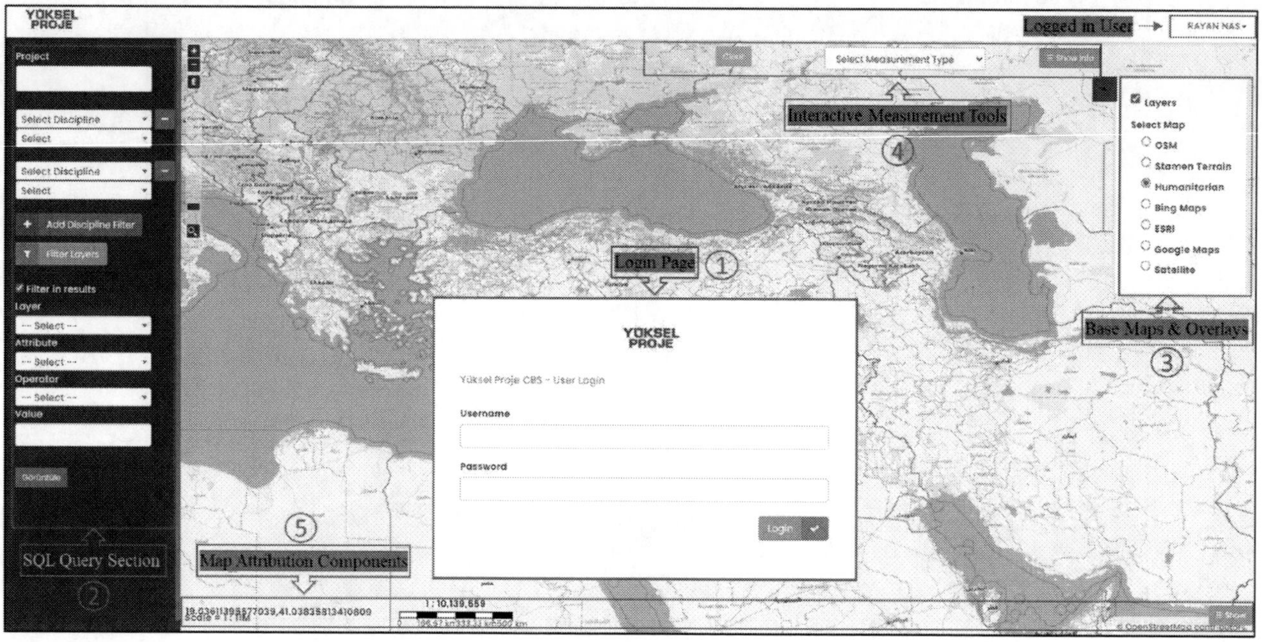

Figure 2. The Web-GIS application interface login screenshot.

queries easily and efficiently, without the need for filtering using expressions. Hence, this component of the Web-GIS application plays a critical role in enabling users to access and analyze the data they need to perform their tasks, making the application a valuable tool for managing and utilizing large amounts of spatial data. The third component (number 3) of the Web-GIS application is where the user can select different base maps and overlays to display on the application interface. This component plays an important role in providing users with a customizable view of the spatial data stored in the geospatial database. Base maps provide the underlying context for the spatial data, and users can choose from a variety of options such as Open Street Map, Google Satellite, and others, depending on their needs and preferences. Overlays, on the other hand, are layers of spatial data that are called from the database and displayed on the map interface, allowing users to visualize and analyze the data in a meaningful way. Examples of overlays could include population density, land use, transportation networks, and others. By providing users with a wide range of base maps and overlay options, the Web-GIS application enables them to view and analyze spatial data in a way that best meets their needs. For example, a user working on a project related to land use may want to use a base map that shows terrain features or vegetation, while also displaying overlays related to land use patterns and zoning regulations. Another user working on a project related to transportation planning may want to use a base map that shows roads and highways, while also displaying overlays related to traffic patterns and transit routes. Thereby, the base maps and overlays management section of the Web-GIS application provides a useful feature and enables users to customize their view of the spatial data, making it easier to analyze and interpret the information they need.

The fourth component (number 4) of the Web-GIS application is an option for measurement tools, which allows users to make interactive measurements on the map interface. This feature enables users to accurately measure distances and areas on the map, providing them with a valuable tool for analyzing and interpreting spatial data. The measurement option in the application allows users to select different measurement types, such as length or area measurements, depending on their needs. For example, a user working on a project related to land use planning may need to accurately measure the area of a particular parcel of land to determine its zoning or permitted land use. The measurement option in the application would enable them to do this quickly and easily, using a variety of measurement tools and techniques. The interactive nature of the measurement option also enables users to make adjustments and refinements to their measurements as needed, ensuring that they are as accurate and precise as possible.

The last component (number 5) of the Web-GIS application that is shown in Figure 2 is the attribution component, which displays information related to the map interface, including the geographic coordinates of the cursor location on the map, the map scale, and the source reference of the map or satellite image (map contributors) being used as the base maps. By displaying this information, the application ensures that users make informed decisions based on the data they are working with. The geographic coordinates provide users with the exact location of their cursor on the map, which can be particularly useful for identifying specific features or locations. The map scale information provides users with an indication of the level

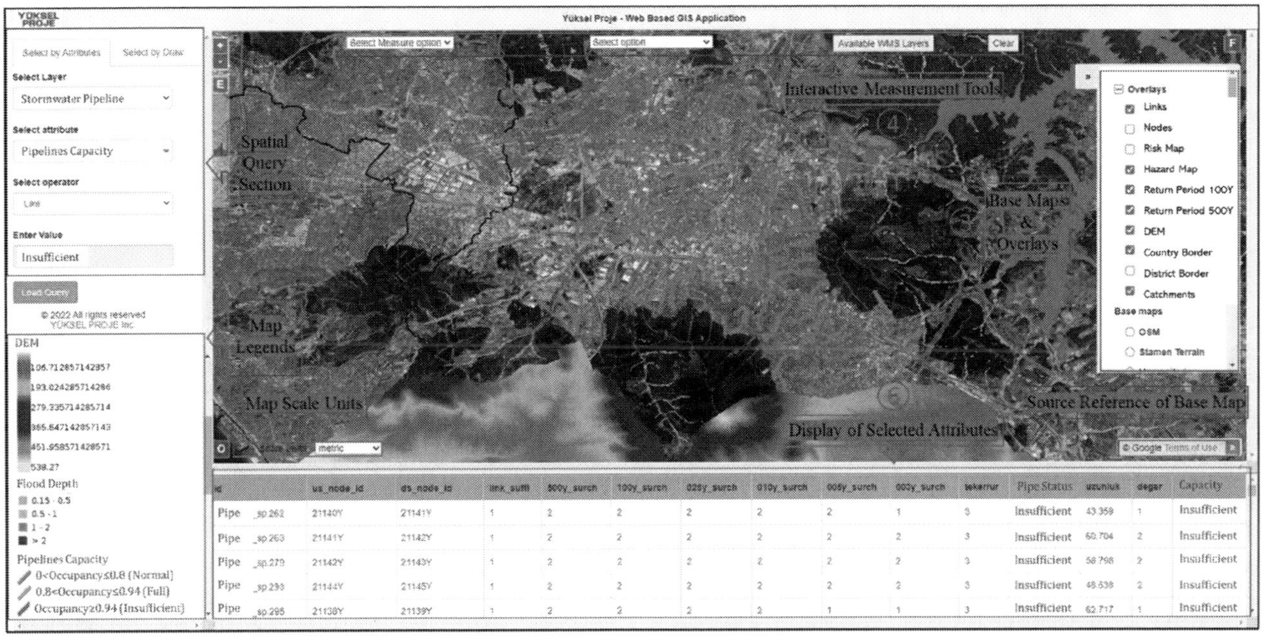

Figure 3. Developed Web-GIS application interface.

of detail being shown on the map, helping them to understand the spatial relationships between different features and elements. Figure 3 presents the developed Web-GIS application's interface, which showcases the functionality and features of the application.

The example data that is called from the database is displayed on the interface, along with the preferences and filters that have been applied to the data. The interface is designed to be user-friendly and intuitive, allowing users to easily navigate and interact with the application. To display the preferred data set on the application interface screen, conditional filtering within the preferred data set can be done with the help of the pre-defined SQL operators as shown in component number 2. The sample opened data in the applications' interface shows whether the capacity of the stormwater pipelines is sufficient or not for a certain rainfall period. The green, orange, and red colors indicate the pipelines' capacities being sufficient, full, and insufficient. The application automatically creates map legends based on the user's data set preferences. The legend appears below Component 2, as shown in Figure 3. Component 6, as shown in the application interface, according to the SQL query made on component 2, displays spatial filtering on the layer based on selected attributes of the data. For instance, in the sample stormwater pipeline data, the user has made a query based on the "Pipeline Capacity" attribute to filter and display the insufficient pipelines. Any data attribute can be selected on this screen, and the selection is dynamically displayed on the map. In addition, as a base map, the user can switch between different base maps, such as satellite images or OpenStreetMap (OSM) invoked via the Internet, as shown in component 3.

III. Discussion

Web-GIS applications have become increasingly popular in recent years, as they provide an easy-to-use and accessible way to process and analyze spatial data. This is particularly useful for decision-makers, who need to quickly access and analyze data in order to make informed decisions about public projects. The Web-GIS application presented in this paper is designed with the needs of a broad audience in mind, making it an ideal tool for users with some level of GIS knowledge. This study presented a Web-GIS application that was designed and implemented with a focus on its features and functionalities. The application allows users to interact with spatial data and utilize its features to explore, analyze, and visualize it. The paper provided an example interface of the application, showcasing its ability to process and display spatial data. The work highlighted the importance of transparency, accountability, and ease-of-use in a Web-GIS application, and emphasized its potential to enable informed decision-making and effective action. In addition, the developed Web-GIS application provides a powerful platform for location-based data analysis, offering various capabilities such as making spatial queries, filtering data, and enabling interactive measurements. These capabilities can be utilized for a wide range of purposes, from environmental monitoring and urban planning to disaster response and emergency management.

One of the key advantages of Web-GIS over traditional desktop GIS is its ease of use. Web-GIS applications are designed to be simple and intuitive, allowing users to quickly and easily process and display geographic data. This makes it an ideal tool for decision-makers who need to take quick measures against certain events, such as floods or other natural disasters. With the Web-GIS application presented in this paper, flood spread maps can be easily classified and visualized for the detection of vulnerable areas according to different flood recurrences, without the need for the complex operations required by traditional desktop GIS. Another advantage of Web-GIS is its ability to enable a large number of users to access and share spatial and non-spatial data within and outside an organization. The Web-GIS application presented in this paper provides global access to data obtained from all disciplines within an institution, making it a valuable tool for collaboration and data sharing. Overall, Web-GIS applications are a powerful and accessible tool for processing, displaying, and analyzing geographic data. Its ease of use and accessibility make it an ideal tool for decision-makers, while its ability to enable data sharing and collaboration makes it a valuable asset for organizations across a range of industries.

Our future work is centered around the development of a dynamic geospatial application that includes several key features, such as a mobile client, 3D Web-GIS, quick and easy query tools, hydrological processing tools, and an integrated BIM and GIS web-based platform. These enhancements are aimed at better meeting the needs of decision-makers and authorities and facilitating collaboration and data sharing within organizations and institutions, particularly those dealing with large datasets. By implementing these improvements, we believe our application will be a valuable tool for promoting effective decision-making and advancing geospatial research.

IV. Conclusion

GIS software is not accessible to everyone, nor spends the time necessary to use it effectively. Desktop GIS is designed for professional users with months of training and experience in GIS. The Web-GIS application developed in this study is based on open source software, including JavaScript, QGIS, PostgreSQL/PostGIS, and GeoServer. This choice of software was made to ensure interoperability between GIS software, database/GIS data entry interface, and map server, respectively, resulting in an efficient and cost-effective Web-GIS system. The Web-GIS application presented in this paper is aimed at a broad audience, primarily for users with some level of experience with GIS programs. In fact, Web-GIS applications are often designed to be simple, intuitive, and convenient, making them easier to use than desktop GIS. Thus, the Web-GIS system becomes a fast and easy way to disseminate the data as a tool for processing, displaying, and analyzing geographic data, while supporting a large number of users at the same time. Therefore, the Web-GIS application is an interesting and powerful tool for decision-makers in designing and planning public projects. Specifically, in order to take quick measures against a certain event, for example, to classify and visualize a flood spread map for the detection of vulnerable areas according to different flood recurrences, the flood spread maps available in the database can be easily displayed on the interface of the application without the need for the operations performed in traditional desktop GIS. In addition, Web-GIS applications enable a large number of users to use services (spatial or non-spatial data) to share with others (global access) within and outside the organization. Therefore, as a result of this work, a web-based GIS application that can be accessed from a computer or mobile device has been designed and presented. This web-based GIS application provides spatial data obtained from all disciplines within an institution and provides quick access to this data when necessary, via the web, regardless of the user's location. Hence, Web-GIS applications have the potential to greatly expand the reach of geospatial analysis and enable a wider range of individuals to engage with geospatial data, fostering greater understanding and informed decision-making in a wide range of fields, from urban planning and environmental management to public health and transportation.

Acknowledgment

The authors would like to thank the colleagues and managers of the R&D and Quality and Management Systems Departments at Yüksel Proje International Co. for their contributions and efforts. The authors would also like to especially thank the management of Yüksel Proje (President and Vice President) and the Coordinator of the Water and Environment Department) for their support and encouragement of scientific and academic research.

References

[1] P. Fu and J. Sun, "Web GIS: principles and applications." Redlands: Esri Press, pp. 89-114, 2011.

[2] R. Nasirzadehdizaji and R. N. Celik, "Open source geo-information technology for making special purpose web-mapping application," Coordinate, 11, pp. 23-26, 2015.

[3] G. Kakaletris, D. Varoutas, D. Katsianis, and T. Sphicopoulos, "Design and Implementation Approaches for Location Based Tourism Related Services" Chapter 18, Geographic Information Systems: Concepts, Methodologies, and Applications, pp. 258-294, 2013.

[4] R. Nasirzadehdizaji, "A web mapping infrastructure design and implementation with open source geo-information technology: A case study of ITU Smart Campus," Master Thesis, Istanbul Technical University, Informatic Institute, Higher Education Council Presidency (YÖK), 2015.

[5] X. M. Chen, "Remote sensing and GIS in environmental risk assessment. Geographic Information Systems: Concepts, Methodologies, and Applications," Chapter 4.15, pp. 840-847, 2013.

[6] M. Ü. Gümüşay, "WebTabanlı Coğrafi Bilgi Sistemi Uygulamaları (YTÜ Davutpaşa Kampüsü) (in English: "Web Based Geographic Information System Applications (Yildiz Technical University Davutpaşa Campus)), Journal of Science and Engineering Sciences, AKU (Afyon Kocatepe University), J. Sci. Eng.17 Special Issue, pp. 215-222, 2017.

[7] X. Wang, Y. Zhu, and L. Yu, "A web-GIS approach for evaluating the effects of green infrastructure on urban thermal environment," Sustainable Cities and Society, vol. 96, p. 103223, 2022.

[8] Y. Zhu, X. Wang, X. Li, and J. Li, "A web-based GIS platform for monitoring and predicting air pollution in urban areas," Environmental Pollution, vol. 287, p. 117395, 2021.

[9] J. Zhang, H. Chen, and Z. Li, "A web-GIS based approach for flood risk assessment and management in urban areas," Natural Hazards, vol. 116, no. 1, pp. 367-383, 2022.

[10] Yüksel Proje International Company. *Web-GIS: User Login.* [Online]. Available from: https://ypnet.yukselproje.com.tr/YpCbs/ 2023.04.05

Development of Data Quality Improvement Method for Hydrodynamic Model of Urban Drainage System Using GIS Capabilities

Cagri Cimen
Department of Water and Environment
Yüksel Proje Inc.
Ankara, 06610, Turkey
e-mail: ccimen@yukselproje.com.tr

Anil Olgac
Department of Water and Environment
Yüksel Proje Inc.
Ankara, 06610, Turkey
e-mail: aolgac@yukselproje.com.tr

Rouhollah Nasirzadehdizaji
Department of Water and Environment
Yüksel Proje Inc.
Ankara, 06610, Turkey
e-mail: rnasirzadeh@yukselproje.com.tr

Abstract—**The pressure on urban drainage systems has increased in recent years due to population growth, urbanization, industrialization, and climate change. The rise in impervious surfaces, such as buildings, roads, and parking lots, as well as airports, ports, and logistics centers, exacerbates the situation. This situation can cause overflows, particularly downstream, as the increase in impervious surfaces leads to shorter periods for surface runoff to be collected and quickly diverted into the drainage system, resulting in large increases in peak flow rates. To plan the existing urban drainage system with current and accurate data using hydrodynamic model tools, it is essential to consider accurate and based on real conditions planning of urban drainage systems and the application of hydrodynamic models that can simplify complex structures into mathematical expressions. In this regard, Geographic Information Systems (GIS) are an effective method for the preparation of various input datasets in accordance with the format and precision required by the hydrodynamic models. This study proposed a methodological approach to prepare critical large-scale data for the hydrodynamic model set up in a GIS environment, aiming to improve the accuracy of analysis results and reflect the real-world situation. Through geospatial analysis and the use of GIS environmental capabilities, key input data, such as Digital Elevation Models (DEMs), land use, buildings' polygons, and catchment areas, were prepared to represent real-world conditions in the model scenarios. The prepared data in the GIS environment provide a detailed and precise representation of complex structures for mathematically realistic simulations. The main finding of this work highlights the importance of using geospatial analysis and GIS tools to prepare input data for hydrodynamic models of urban drainage systems, leading to improved accuracy and efficiency of urban drainage services.**

Keywords—Geographic Information System; sustainable urban drainage; urbanization; stormwater system; climate change.

I. Introduction

Accurate data that reflects real-world conditions play an important role in planning sustainable urban drainage systems. In metropolitan areas, population growth and unexpected urban development lead to increasing challenges, difficulties, or insufficient planning to meet the future demands of urban infrastructure services, such as urban drainage systems [1] [2] [3]. In line with this, the existing urban drainage systems need to be improved to comprehensively utilize the simulations of different complex and large-dimensional data to incorporate all the dynamics of the real situation.

Improvements of the physical data sets of the existing stormwater system (ground elevation, flow elevation, size, route, etc. of pipe and manhole system elements) and creeks (route, cross-section, profile, etc.) prior to the integration of these data into the hydrodynamic model are the subject of much more detailed studies. To this end, with engineering approaches the GIS environment is an efficient system for the preparation of input datasets in accordance with the format and precision required by the hydrodynamic models. With the improvement of the quality of input data and incorporation of hydrological information, hydraulic analysis of urban drainage systems can be efficiently performed. However, such analyses are generally performed on models of systems that are independent of each other. The hydrodynamic model and its hydraulic analysis are performed either for the Stormwater Collection System (SWCS) alone or for the creeks individually, and the system capacity and its dimensions are determined by the rainfall data used.

The hydraulic analyses may not always reflect the interaction with time according to the existing conditions in the model results due to the site characteristics. Accordingly, the analysis results either encounter over-design upstream and downstream of the system or insufficient sizing in the downstream portion. Furthermore, since the evolution of downstream conditions over time cannot be clearly defined in non-integrated models, it is difficult to obtain holistic simulations. However, in 2-Dimensional (2D) models, time-dependent and dynamic solutions of integrated holistic models of the stormwater collection system and creeks, the relationship of the systems with their environment can be analyzed in terms of time variables, especially by defining additional data reflecting real conditions to the models. In this way, mainly at the points where the systems are connected to each other (i.e., junctions), and towards the upstream or downstream impacts can be analyzed closer to the actual situation. Hence, the improvement of the data to quality and accuracy that can be integrated into the model, especially focusing on the required data for the hydrodynamic models, ensures the model's precision and effectiveness in reflecting the real condition. The important data include the physical data for stormwater and creeks, as well as hydrological inputs that are the main data for any 1D or 2D hydraulic models. The DEM with a certain precision, land use data including geometry and land use types, buildings' polygon, and surface water catchment areas are fundamental data for hydrodynamic models and thus for the efficient urban drainage system design and implementation.

The DEM data, which digitally represents the earth's surface topography, is used for many different purposes. One of the applications of DEM data is to create hydraulic or hydrodynamic models. The analysis of these models' results leads to obtaining different outputs for monitoring and investigation of the hydraulic and hydrologic process of the system, such as pipeline capacities, flood maps (flood extensions, depth, and velocity), etc. Hence, terrain data plays an important role in determining the accuracy of floodplains. Different methods are available for creating and obtaining the DEM data. One of them is Light Detection And Ranging (LiDAR) to make high-resolution maps for digital 3-D representations of areas on the earth's surface. It has been stated by various researchers that the floodplain maps derived from this dataset give much more accurate results than the maps obtained using other available topographic datasets in the world due to their very high resolution [4] [5] [6].

Land use maps, on the other hand, contain different information related to land, such as land use types and geometries, and are mainly obtained from the Coordination of Environmental Information (CORINE) land cover maps [7]. Land use maps provide geographic information on land cover, including land use/cover characteristics and changes, vegetation state, water cycle, and earth's surface energy variables, to a wide range of users in environmental terrestrial applications. In addition to these maps, some numerical boundaries should be integrated into the model in order to define the existing conditions in the hydrodynamic model.

The objective of this study was to enhance the accuracy and efficiency of urban drainage services by utilizing geospatial analysis and GIS tools to prepare input data for hydrodynamic models of urban drainage systems. Specifically, the study proposed a methodological approach to improve the physical data sets of the existing stormwater system and creeks prior to their integration into the hydrodynamic model. The remainder of the paper is structured as follows: Section II details the methodology employed to enhance the physical data, including its integration into the hydrodynamic model and preparation in the GIS environment. Section III highlights the significance and contributions of this work in improving hydrodynamic models for urban drainage systems, resulting in the provision of efficient and adequate urban drainage services.

II. METHODOLOGY

The impact of surface runoff on urban areas can be reduced through the implementation of Low Impact Development (LID) techniques and a Sustainable Urban Drainage System (SUDs), which integrate traditional drainage systems with LID methods. This approach will make urban areas more resilient and sustainable in the face of climate change and contribute to Water Sensitive City (WSC) targets and Integrated Urban Water Management (IUWM). Monitoring and evaluating the effectiveness of the SUDs is crucial for ensuring their long-term success. The preparation of input data for a hydrodynamic model using GIS capabilities involves several engineering approaches that aim to ensure the continuity of physical datasets from upstream to downstream. These methods resolve stability issues that may arise from using DEMs, prepare relevant hydrological and hydraulic parameters, such as rainfall patterns and runoff coefficients, and create design data, such as rainfall catchments. Hence, these techniques enable the hydrodynamic model to run smoothly and provide accurate results. Therefore, the focus of this study was to develop a method for improving the data quality of hydrodynamic models of urban drainage systems using GIS capabilities. The hydraulic modeling process for the assessment of the adequacy of existing stormwater collection systems and creeks includes the following steps.

- Transferring the physical data of the urban drainage system, including the stormwater collectors and creeks, which have been improved and of sufficient quality, to use in the hydraulic model.
- Definition of initial condition boundaries for each hydrological scenario.
- Definition of surface flow coefficients.
- Defining the surface roughness coefficient.
- Identification of the surface model (i.e., DEM).
- Definition of the land use map.
- Defining buildings' polygon to the model.
- Identification of sub-basins to the hydraulic program.
- Running the hydraulic model of the relevant basin.

In general, urban drainage system analyses are performed by integrating the data of the system into the model and defining the hydrological/hydraulic inputs. Whether the physical data for the system is obtained from relevant institutions or generated from classified data, these data are necessary to improve the quality and convert it into the appropriate format for entry into the hydrodynamic model. In the improvement of data quality and to ensure upstream-downstream continuity, the information of all system elements from the upstream of the system to the discharge point is transferred to the database. The next step in the setup of the model is the introduction of hydrologic data into the model, identifying initial conditions, such as rainfall intensity, and the collection time (time of concentration). By introducing surface discharge and roughness coefficients into the model as hydraulic parameters, the model setup process is completed at a level sufficient for a general analysis of the system. However, more precise planning is needed because of the global trend of planning for the WSC targets through the SUDs, an important pillar of IUWM, and the LID framework. Accordingly, for a sustainable urban drainage system, the actual conditions need to be defined mathematically in order to analyze the hydrodynamic model of the system more precisely. These inputs containing real conditions are primarily DEM data and land use maps, buildings' polygon, and catchment areas for the stormwater collection systems.

LiDAR data, which is used as a digital elevation model and includes precise elevation information of the earth's surface needed is important data for 3-D modeling the terrain features, such as buildings, transportation, flow channels, etc. The use of LiDAR data is to carry out accurate engineering calculations along with the analysis of the current situation, to plan the urban drainage system. For the analysis and planning of urban drainage systems, in addition to the surface digital model, land use data including land use type and geometric shape and their attribute information is required in the hydrodynamic model. The land use data provide the possibility to evaluate the results of the analysis based on the land utilization of the existing system. In order to define the actual condition of the terrain in the model, buildings' polygons representing the densest structures need to be integrated into the surface model along with land use data. In addition, the sub-basins, which form the accumulation areas of the superficial flow of rainfall, need also to be introduced into the model. The urban drainage system areas covered by rainfall and resulting runoff are defined in the model to accurately calculate the model's flow and to allow the model to identify the actual flow from the defined precipitation rate. The workflow schema for setting up and running the model is shown in Figure 1. The workflow diagram outlines the process from model setup to model runtime, and shows the datasets defined and integrated into the model.

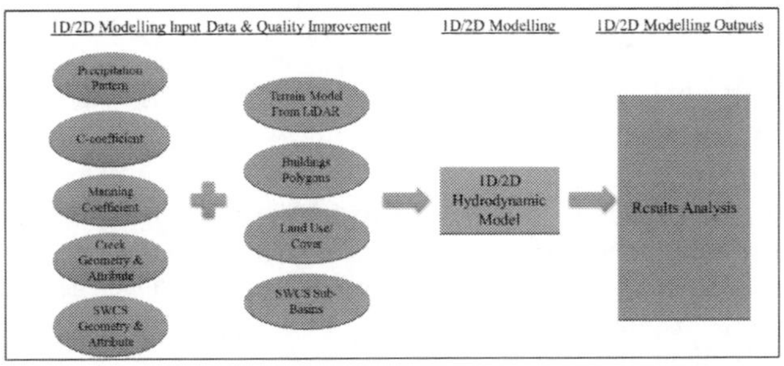

Figure 1. Workflow diagram for setting up and running a hydrodynamic model of an urban drainage system.

In the first stage of the urban drainage system modeling setup, the main inputs are rainfall patterns, runoff and roughness coefficients, and the physical data of the system, as indicated by the green clusters. For more accurate modeling results as specified earlier, the inputs that reveal the actual situation are shown in cyan color clusters (second stage) and include LiDAR, buildings' polygon, land use, and sub-basin areas. After the model setup, the process of running the hydrodynamic model includes both the optimizations in associating the model with the input data and the control and validation of the process steps in the input-run-outcome process (third stage). After all arrangements and controls, the results of the analysis are captured and interpreted for the urban drainage system planning (fourth stage). The procedures for preparing input data in a GIS environment prior to hydrodynamic model installation and processes are discussed in detail in the following sub-sections.

This study has performed the GIS operations and geospatial processing and analysis with the potential use of the open-source Quantum Geographic Information System (QGIS) [8] software.

A. Digital Elevation Model

In general, a DEM derived from any remote sensing data source represents the elevation of the bare ground surface, excluding features, such as buildings and trees. This is because the technology used to generate DEMs is primarily designed to capture elevation information of the ground surface, rather than the elevation of other features above the ground. However, with LiDAR data, it is possible to generate a DSM (Digital Surface Model) that includes the elevation of all features on the surface, including buildings and trees. The DSM can be derived from the same LiDAR data that is used to generate the DEM, by including all the points in the LiDAR point cloud, rather than just the ground points. It is important to note that while the DSM provides a more comprehensive representation of the surface, it may not be suitable for some applications that require the elevation of only the bare ground, such as hydrological or hydraulic modeling. In these cases, the DEM derived from LiDAR data needs to be modified and improved with the applied techniques.

Although the LiDAR data with a resolution of 0.25x0.25 meters provided in the study was perfectly well for the hydraulic model, however, this data was not used in both hardware and temporal terms due to its need for a very large memory capacity. Hence, the resolution of this data has been used for the urban area with a resolution of 2.00x2.00 m by various optimization processes. There have also been several procedures to reflect creeks on the surface model to calculate the superficial flow. For processing the rehabilitated creeks to the DEM, by using the required items (i.e., the starting end codes, cross-sections, and dimensions), the rehabilitated creeks' base and its slop guidelines (i.e., top and upper boundary lines) are defined according to the width, slop inclination and cross-section types. The height values of the points in the creek base are calculated along the creek, depending on the slope of the stream, by creating a point every 1 m on the lines. The values of the points on the upper slope and boundary lines on the slope of the creeks were calculated by determining the elevation at the relevant location, taking into account the base slope of the creek. LiDAR data was primarily converted into a point data type for each basin. Then, the points were deleted within the region up to the threshold value (50 cm) outside the outer part of the rehabilitated creeks (on-slope boundary lines). Points excluding these removal points are merged with the set of points created for the rehabilitated creek. As shown in Figure 2 the merged file is the final data to be used as a modified surface model to input to the hydrodynamic model.

B. Land Use Map

An assessment of an urban drainage system by overlapping an existing land-use map is a priority process. There are critical assessments of the impact of drainage systems

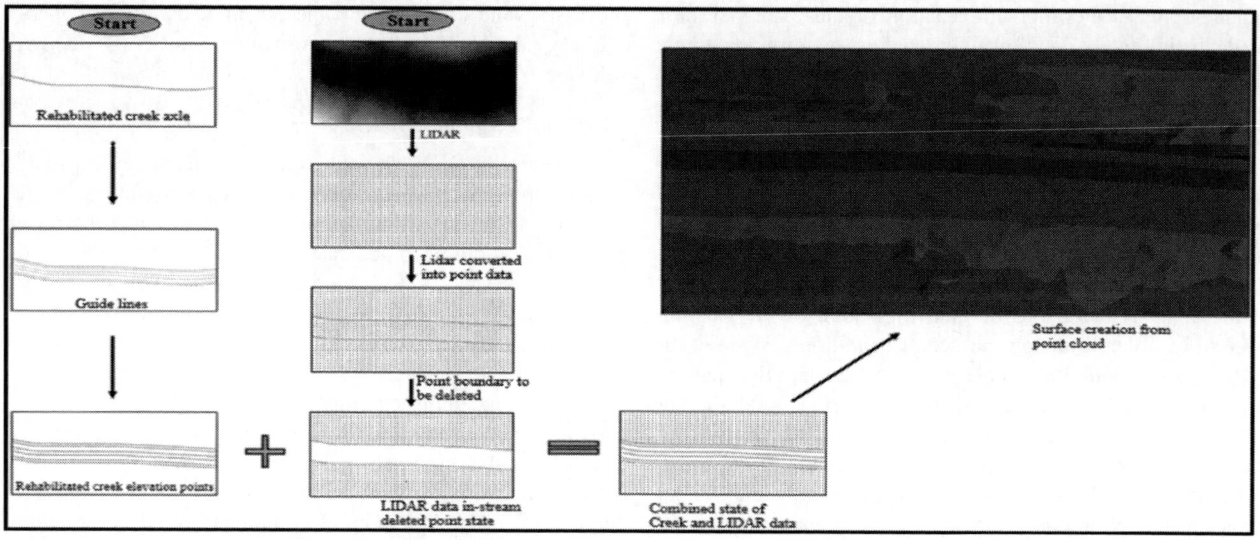

Figure 2. Surface model modification processes.

overlapping with land use, such as different types of settlements, industrial-commercial-public areas, infrastructure facilities, such as roads-ports-airports, and zoning plans involving different types of uses, such as agricultural and forest zones. In this study, the CORINE land cover map is used as the land use data. The analysis of how the extent of the land use boundary including the existing urban drainage system or the impact on the land use type in the model has been conducted through the GIS.

C. Buildings' Polygon

In the improvement process of buildings' polygons, which is an important input to the model, topological analysis is performed first. In these analyses, the "must not overlap", "must not have gaps", and "must not have duplicates" rules were applied. Without these corrections, the meshing process produces a large number of triangles formed in small areas of geometrically and hydraulically insignificant dimensions, including small areas that cannot be detected easily. Hence, it maximizes the meshing time and modeling time, distorts modeling stability, and may even cause the model to fail. There is no existing ready-made GIS application or plugin to eliminate errors detected in the analysis. For this reason, the "Aggregate", "Simplification" and "Deleting Holes" functions were used in GIS for the buildings' polygon dataset. The detailed specifications of the "Aggregate", "Simplification" and "Deleting Holes" functions for buildings' polygons are provided below.

Buildings that are too close together cause small triangles during the meshing process. In the experiments, it has been seen that it is necessary to simplify the buildings closer than 2 m, which do not cross streets, into a single polygon. To this end, different GIS methods have been applied, however, the "Aggregate Polygons" process has resulted in favorable results. In order to convert the shortest side of the building into two or three cells, the cell size should be small enough. The values of the input cells encompassed by the coarser output cells are aggregated by one of the maximum, minimum, mean, median, and sum statistics for the input cells' value. No data values were ignored by the aggregation calculation.

The "Simplification of polygons" as the generalization operation of the boundaries was used to remove the extraneous bends and small intrusions and extrusions from the buildings' polygon boundaries without destroying their essential shape. The "Point remove" and "Bend simplify" algorithms were applied for non-orthogonal polygon boundaries. The "Point remove" is efficient for data compression and for eliminating redundant details; however, the line that results may contain sharp angles and spikes that reduce the cartographic quality of the line. The "Point remove" algorithm is used for compression or small amounts of data reduction and when high cartographic quality is not needed. The "Bend simplify" was applied for shape recognition to detect bends, analyze their characteristics, and eliminate insignificant ones. In this method, a bend that's too narrow is widened slightly to satisfy the tolerance. The resulting line is more faithful to the original and provides better cartographic quality. In order to achieve the minimum triangle size of the mesh, it needs to reduce the number of fractures on the buildings' polygon without distorting their shape. Simplifying buildings' polygon is used for reducing the detail at the boundaries of buildings while maintaining the basic shape and size of the buildings. Buildings are usually rectangular areas; therefore, the simplification process preserves and maximizes rectangularity. Each individual building itself has been simplified. Buildings connected by straight lines were not simplified as a group but on the whole boundary of a building. Thereby, this method describes buildings as topologically discrete, connected by straight lines close to parallel to each other, and connected by more complex paths.

The boundaries of discrete buildings or buildings connected by straight lines were modified so that all angles close to 90 degrees still remained exactly 90 degrees. Some edges were smoothed, reducing the number of fractures, but the areas remain roughly the same as the original [9]. The maximum degree of simplification is reached when a building is reduced to a quadrilateral. Once the simplification tolerance is relatively large relative to the size of the building, the building is simplified directly to a rectangle centered on its center of gravity, but the area remains the same. The sides of the resulting rectangle are in the same ratio as the sides of the bounding box aligned to the longest side of the original building.

Areas between buildings or adjacent building blocks are considered as the terrain surface in the model and if these gaps are not cleared before being imported to the model, during the meshing, the model fills these gaps with triangles. accordingly, these areas are going to fill up like puddles during modeling. In order to avoid this problem, it has been accepted to include these spaces as building areas. For this process, the holes were deleted or filled using a GIS operator called "Deleting holes". This operator generates a new output feature class containing the features from the input polygons with some parts or holes of a specified size deleted and exploded into single parts. The "Deleting Holes" operator combines both of the above functionalities which can optionally specify a maximum area of hole to delete. In this algorithm, the maximum area parameter refers to the maximum size of the hole to be filled. Hence, an optional minimum area parameter allows only holes smaller than a certain area threshold to be removed. For instance, leaving this parameter at 0.0 the algorithm removes all hole sizes. An example of the described generalization of buildings' polygon is shown in Figure 3.

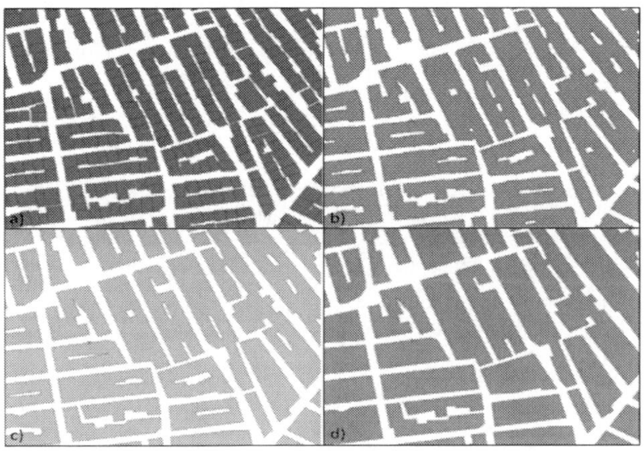

Figure 3. Illustration of used GIS operations to buildings' polygon data quality improvements and their results for the sample dataset. (a) Original data, (b) the aggregated polygons, (c) simplified polygons, and (d) deleted holes from the buildings' polygon.

Therefore, the created new output data created from the outer boundaries contained polygons of filled holes with preserved attributes of the data. The resulting polygons were then manually checked and loaded into the hydrodynamic model.

D. *Water Collection Catchments*

In a 2D hydrodynamic model for analysis of an urban drainage system involving major stormwater collection lines and creeks, modeling can be performed on a macro-scale with only the terrain model and urban drainage systems without additional definitions of terrain specifications. However, the use of large basins in 2d modeling to avoid the impact of minor network pipes not entering the model, and therefore, using large basins as sub-basins in 2d modeling in the planning of a micro-scale urban drainage system provides more detailed and realistic results. For this, different tools and plugins are used (e.g., SWAT, SWAT+, ArcGIS Hydrographic Tools, SAGA Terrain Analysis- Hydrology, etc.) prior to running the hydrodynamic model.

The delineation of sub-basins is a time-intensive process that requires a high level of computing capacity due to the size of the surface model data and the abundance of components. Figure 4 shows the terrain surface model improved from LiDAR data (right) and the system's physical data defined in the hydrodynamic model including stormwater collection systems (i.e., pipelines networks) and delineated sub-basins and collection basins by Voronoi (Thiessen) polygon's method in the GIS environment (left).

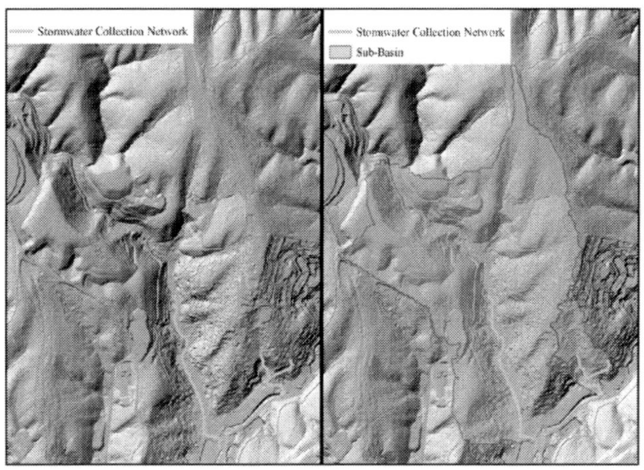

Figure 4. Example of a sub-basin delineated with SWAT+ plugin.

In this study, the delineation of sub-basin boundaries was mainly carried out on QGIS, an open-source GIS software, and through its compatible hydrological processing and analysis plugin called SWAT+ [9]. In the delineation process, the improved DEM data (i.e., eliminated above-ground features) derived from LiDAR data was used as the surface model. In the sub-basin delineation procedure, the outlets and discharge points, the definition of upstream-downstream continuity in physical data, and the stormwater collection system (the program considers the stormwater collection system as a streamline) are also introduced to the model allowing the program to establish sub-basin boundaries accordingly. In order to distribute the incoming flow to the manholes, the delineated basins were divided also into collection basins of manholes by Voronoi (Thiessen) polygon's method which is also available in QGIS [8].

E. Mesh Corrections Based on the Buildings' Geometry

Buildings are not located within stormwater collection systems and the building's polygons have been excluded from the calculation since the building's polygons are defined as void and without meshing as shown in Figure 5 (left). Hence, the buildings' geometries needed to be defined in the model as a component of the stormwater collection systems. In other words, each building's block is introduced into the model using GIS as a discrete catchment. Additionally, to transfer the roof rain waters to the surface mesh, as shown in Figure 5 (right), the buildings' roof drainage points are defined and the outlet flow elevations were adjusted to form a 1/100 slope depending on the mesh elevation code. Therefore, by using this method the buildings' polygons have been defined to be considered in the hydrodynamic analysis. Mesh Corrections Based on the buildings' geometry and definition of buildings' roof drainage points to the model. Figure 5 shows the building's polygons as void and without meshing (left), and the definition of buildings' geometries in the model as the component of the stormwater collection systems (right).

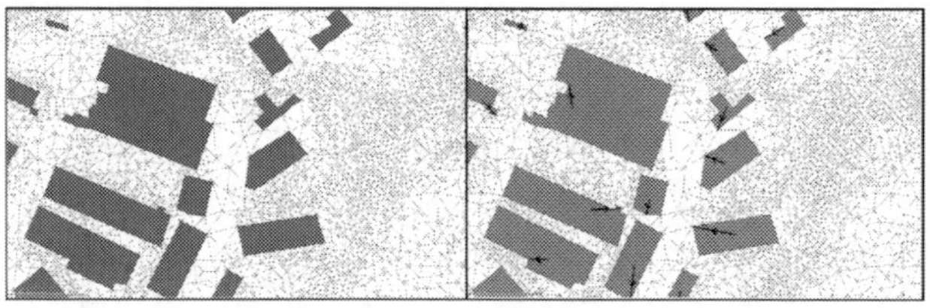

Figure 5. Mesh corrections for buildings' geometries.

The symbols with black lines and arrows in Figure 5 (left) are indicating the buildings' roof water drainage points on each building's block, and the outlet flow to transfer the roof rain waters to the closest surface mesh.

III. CONCLUSION

The achievement of long-term planning and sustainable stormwater management systems in terms of environmental, and economic to ensure adequate and efficient urban drainage services is important in order to adapt to today's fast-growing conditions. In terms of sustainable management practices, precise planning with accurate and actual data that represents reality needs to be targeted throughout the engineering, administrative and economic measures for the continuity of service quality in which investment and operation strategies are in accordance with the smart city concept. To meet the needs of urban drainage systems to adapt to the predicted circumstances and changes that vary depending on the population growth trend, fast development of cities, and climate change issues, It is appropriate to conduct an analysis of the actual situation of the system. In this framework, it is necessary to use a hydrodynamic simulation to take into account integrated and dynamic conditions, as well as to ensure that the existing system's data are correctly and appropriately defined in the model. This study presented a methodological approach to the preparation of important large-dimensional data in the setup of a hydrodynamic model to reflect the actual conditions and achieve more precise analysis results. In this regard, the high potential use of GIS and its capabilities as well as geospatial processing and analysis in a GIS environment play an important role in performing data preparation and quality improvement. Therefore, the development of a methodological approach for the preparation of accurate data as the input of the hydrodynamic models of urban drainage systems, using GIS capabilities ensures adequate and efficient urban drainage services to adapt to the fast-growing conditions in a metropolitan city.

ACKNOWLEDGMENT

The authors would like to thank the colleagues and managers of the R&D and Quality and Management Systems Departments at Yüksel Proje International Co. for their contributions and efforts. The authors would also like to especially thank the management of Yüksel Proje (President and Vice President) and the Coordinator of the Water and Environment Department) for their support and encouragement of scientific and academic research.

REFERENCES

[1] Z. B. Uygun, "Investigation of Stormwater Collection System Problems in İstanbul," İstanbul Technical University, Graduate School of Science Engineering And Technology, November 2019.

[2] EPA, Environmental Protection Agency, "Reducing Stormwater Costs through LID Strategies and Practices," 841-F-07-006, 2007.

[3] G. Ferrier, D. Milan, C. Keat Yew, and R. J. Pope, "Potential of Retrofitting Sustainable Urban Drainage Systems Using an Integrated Geographical Information System-Remote Sensing Based Approach," Department of Geography, Environment and Earth Sciences, University of Hull, Hull, HU6 7RX, UK, 2018.

[4] A. Casas, G. Benito, V. R. Thorndycraft, and M. Rico, "The topographic data source of digital terrain models as a key element in the accuracy of hydraulic flood modelling," Earth Surface Processes and Landforms Volume 31, Issue 4, 2006.

[5] G. Schumann, et al., "Comparison of remotely sensed water stages from LiDAR, topographic contours and SRTM,", ISPRS Journal of Photogrammetry and Remote Sensing, 2008.

[6] M. E. Charlton, A. R. G. Large, and I. C. Fuller, "Application of airborne LiDAR in River environments: The River Coquet, Northumberland, UK,", Earth Surface Processes and Landforms, 2003.

[7] CORINE Land Cover maps, European Environment Agency (EEA), Copenhagen, Denmark, 2018.

[8] Swiss Society of Cartography, "Cartographic Generalization—Topographic Maps," Cartographic Publication Series, No 2, 1987.

[9] K. Bieger, et al., "Introduction to SWAT+, a completely restructured version of the Soil and Water Assessment Tool." JAWRA Journal of the American Water Resources Association, 53(1): pp. 115–130, 2017.

Graph Neural Network for Accurate and Low-complexity SAR ATR

Bingyi Zhang*, Sasindu Wijeratne*, Rajgopal Kannan†, Viktor Prasanna*, Carl Busart†
*University of Southern California †DEVCOM US Army Research Lab
e-mail: *{bingyizh, kangaram, prasanna}@usc.edu †{rajgopal.kannan.civ, carl.e.busart.civ}@army.mil

***Abstract*—Synthetic Aperture Radar (SAR) Automatic Target Recognition (ATR) is the key technique for remote sensing image recognition. The state-of-the-art works exploit the deep convolutional neural networks (CNNs) for SAR ATR, leading to high computation costs. These deep CNN models are unsuitable to be deployed on resource-limited platforms. In this work, we propose a graph neural network (GNN) model to achieve accurate and low-latency SAR ATR. We transform the input SAR image into the graph representation. The proposed GNN model consists of a stack of GNN layers that operates on the input graph to perform target classification. Unlike the state-of-the-art CNNs, which need heavy convolution operations, the proposed GNN model has low computation complexity and achieves comparable high accuracy. The GNN-based approach enables our proposed *input pruning* strategy. By filtering out the irrelevant vertices in the input graph, we can reduce the computation complexity. Moreover, we propose the *model pruning* strategy to sparsify the model weight matrices which further reduces the computation complexity. We evaluate the proposed GNN model on the MSTAR dataset and ship discrimination dataset. The evaluation results show that the proposed GNN model achieves 99.38% and 99.7% classification accuracy on the above two datasets, respectively. The proposed pruning strategies can prune 98.6% input vertices and 97% weight entries with negligible accuracy loss. Compared with the state-of-the-art CNNs, the proposed GNN model has only 1/3000 computation cost and 1/80 model size.**

Keywords*—Synthetic aperture radar, automatic target recognition, graph neural network, low computation complexity, model pruning

I. Introduction

Synthetic aperture radar (SAR) is capable of high-resolution remote sensing and independent of weather conditions to observe the targets on the earth ground. SAR automatic target recognition (ATR) is the crucial technique to classify the target in the SAR images and has been used in many real-world applications, such as agriculture [1] [2], civilization [3] [4], etc. SAR devices are typically mounted on moving platforms, such as aircraft, spacecraft, and small/micro satellites [5]–[9]. These moving platforms usually have limited computation resources and power budgets (e.g., 80-180W [10]). The state-of-the-art works [11]–[15] develop complex convolutional neural networks (CNNs) for SAR ATR to achieve high classification accuracy. However, complex CNNs suffer from high computation costs and large memory footprints, making them unsuitable to be deployed on resource-limited platforms. For example, to achieve real-time image classification using CNNs, GPU is widely used. The power consumption of a state-of-the-art GPU device (e.g., NVIDIA RTX3090 has a power consumption of 450W) can exceed the power budget of the small/micro satellites.

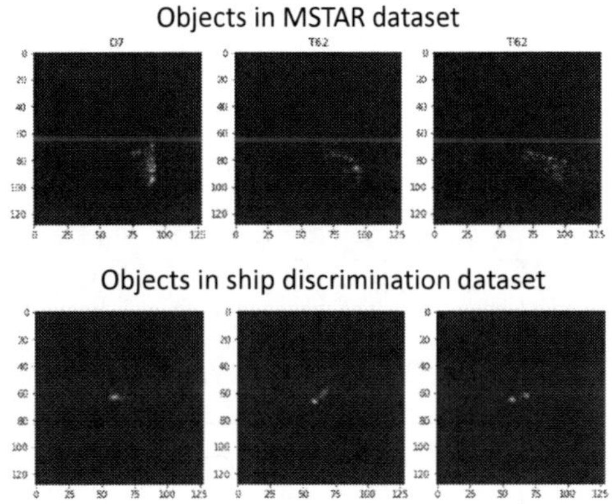

Figure 1. The objects in the SAR images

We identify that CNNs have high computation costs due to (1) heavy convolution operations and (2) CNNs do not exploit the data sparsity in SAR images because CNNs need to use the whole image as input. As shown in Figure 1, an object in a SAR image usually has a small number of pixels, and most pixels are irrelevant for classification. Recently, Graph Neural Networks (GNNs) are proposed to operate on graph data structure and have been successfully applied to many graph classification tasks [16]–[18], such as point cloud classification. [19] has proven that GNN can classify a graph based on its graph structural information and vertex features. Motivated by that, we propose to use GNN for SAR ATR. First, we extract the image pixels of the target object. We use these pixels to build a graph by constructing the edge connections among the pixels. We exploit GNN to operate on the input graph for target classifying. The proposed GNN-based approach achieves significantly less computation cost and comparable accuracy compared with state-of-the-art CNNs. Moreover, we propose attention mechanisms, including vertex attention and feature attention, to improve the model's accuracy. Our main contributions are:

- We propose a novel GNN model for SAR ATR with attention mechanisms, including vertex attention and feature attention, to achieve high accuracy with low computation complexity.
- We propose the input pruning strategy and the weight pruning strategy to further reduce the computation complexity with negligible accuracy loss.
- We perform detailed ablation studies to evaluate (1) various connectivity for constructing the input graph, (2) various types of GNN layers, (3) the effect of the attention

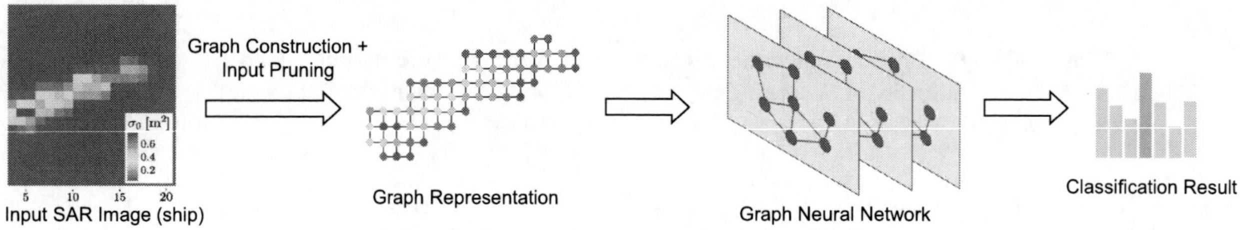

Figure 2. Overview of the proposed approach

TABLE I. NOTATIONS

Notation	Description	Notation	Description
$\mathcal{G}(\mathcal{V},\mathcal{E},\boldsymbol{X}^0)$	input graph	v_i	i^{th} vertex
$\mathcal{V}$	set of vertices	e_{ij}	edge from v_i to v_j
$\mathcal{E}$	set of edges	L	number of GNN layers
$\boldsymbol{h}_i^l$	feature vector of v_i at layer l	$\mathcal{N}(i)$	neighbors of v_i

mechanism, and (4) the impact of the proposed pruning strategies.

- We evaluate the proposed approach on MSTAR and ship discrimination datasets. The evaluation results show that the proposed GNN model achieves 99.38% and 99.7% classification accuracy on the above two datasets, respectively. Compared with the state-of-the-art CNNs, the proposed GNN model has only 1/3000 computation cost and 1/80 model size.

The rest of the paper is organized as follows: Section II presents the proposed GNN model for SAR ATR; Section III describes the proposed pruning strategies for reducing computation complexity; Section IV demonstrates the evaluation results.

II. Proposed Model

Figure 2 depicts the overview of the proposed approach. In Section II-A, we introduce the basics of the graph neural network. In Section II-B, we cover the proposed graph representation for the SAR images. In Section II-C, we introduce the proposed GNN model architecture.

A. Graph Neural Network

We define GNN notations in Table I. Graph Neural Networks (GNNs) [20]–[22] are proposed for representation learning on graph $\mathcal{G}(\mathcal{V},\mathcal{E},\boldsymbol{X}^0)$. GNNs can learn from the structural information and vertex features and embed this information into low-dimension vector representation/graph embedding (For example, $\boldsymbol{h}_i^L$ is the embedding of vertex v_i). The vector representation can be used for many downstream tasks, such as node classification [21] [20], link prediction [23], graph classification [24], etc. As shown in Figure 3, GNNs follow the message-passing paradigm that vertices recursively aggregate information from the neighbors.

Input: Graph: $\mathcal{G}(\mathcal{V},\mathcal{E})$; vertex features: $\left\{\boldsymbol{h}_1^0,\boldsymbol{h}_2^0,...,\boldsymbol{h}_{|\mathcal{V}|}^0\right\}$;
Output: Output vertex features $\left\{\boldsymbol{h}_1^L,\boldsymbol{h}_2^L,...,\boldsymbol{h}_{|\mathcal{V}|}^L\right\}$;

for $l = 1...L$ **do**
  for each vertex $v \in \mathcal{V}$ **do**
    $\boldsymbol{a}_v^l = \text{Aggregate}(\boldsymbol{h}_u^{l-1} : u \in \mathcal{N}(v))$
    $\boldsymbol{z}_v^l = \text{Update}(\boldsymbol{a}_v^l, \boldsymbol{W}^l)$, $\boldsymbol{h}_v^l = \sigma(\boldsymbol{z}_v^l)$

Figure 3. GNN Computation Abstraction

B. Graph Representation

We transform the input SAR image into a graph representation $\mathcal{G}(\mathcal{V},\mathcal{E},\boldsymbol{X}^0)$, where each pixel in the SAR image is mapped to a vertex $v \in \mathcal{V}$ in the graph. The SAR signal value of the pixel becomes the feature of the vertex. Each pixel is connected to its neighbors as the edge connections $\mathcal{E}$. As shown in Figure 4, we propose the following two ways of connecting a pixel to its neighbors and evaluate them in the experiments:

- **4-connectivity**: Each pixel is connected to the four neighbors: up ($p2$), down ($p8$), left ($p4$), and right ($p6$).
- **8-connectivity**: Each pixel is connected to the eight neighbors: $p1$, $p2$, $p3$, $p4$, $p6$, $p7$, $p8$, $p9$.

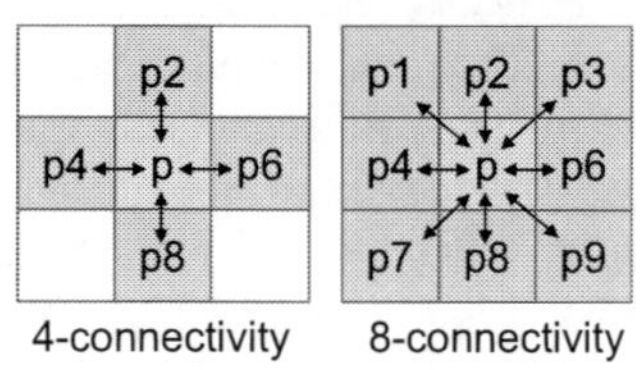

Figure 4. Two types of connectivity for constructing input graph

C. Model Architecture

The proposed model architecture is shown in Figure 5, which consists of a stack of layers, including Graph Neural Network layers, graph pooling layers, and attention layers. The final Multi-layer Perceptron (MLP) generates the classification result. For simplicity, $v_{i,j}$ denotes the vertex/pixel that locates at i^{th} row and j^{th} column in original SAR image. The input to layer l $(1 \leqslant l \leqslant L)$ is the vertex feature vectors $\{\boldsymbol{h}_{i,j}^{l-1} : v_{i,j} \in \mathcal{V}_{l-1}\}$ and edges $\{e : e \in \mathcal{E}_{l-1}\}$ that defines

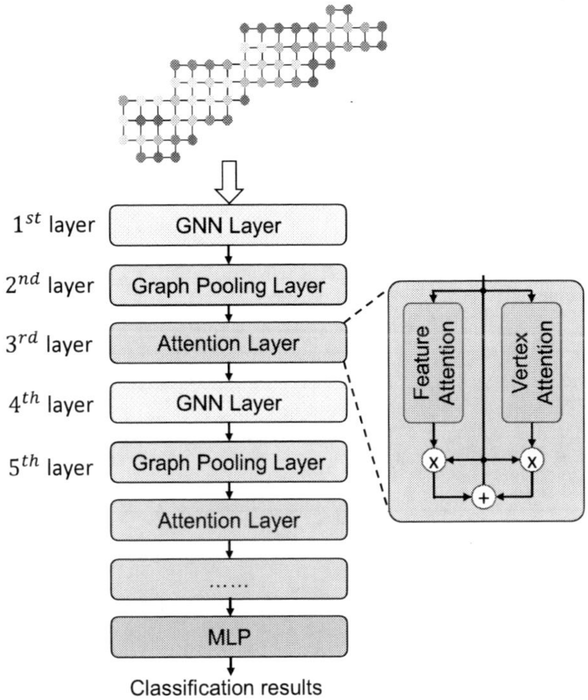

Figure 5. Diagram of model architecture

the connectivity of the vertices in $\mathcal{V}_{l-1}$. The output of layer l is the vertex feature vectors $\{\boldsymbol{h}_{i,j}^{l} : v_{i,j} \in \mathcal{V}_l\}$.

Graph neural network (GNN) layer: A GNN layer follows the *Aggregate-Update* paradigm as shown in Algorithm 3. Using the Aggregate() function, each vertex aggregates the feature vectors from the neighbors (line 3 of Algorithm 3). Then, each feature vector is updated by the Update() function to generate the updated feature vector (line 4 of Algorithm 3). There are some representative Graph Neural Network layers, such as GCN [20], GraphSAGE [21], GIN [19], and SGC [25].

Graph pooling layer: It downscales the input graph $\mathcal{V}_{l-1}$ into a smaller output graph $\mathcal{V}_l$. The pooling operaton is similar to the pooling in the 2-D images:

$$\boldsymbol{h}_{i,j}^{l} = \max(\boldsymbol{h}_{2i,2j}^{l-1}, \boldsymbol{h}_{2i+1,2j}^{l-1}, \boldsymbol{h}_{2i,2j+1}^{l-1}, \boldsymbol{h}_{2i+1,2j+1}^{l-1}) \quad (1)$$

where $v_{i,j}^{l} \in \mathcal{V}_l$, and $v_{2i,2j}^{l-1}, v_{2i+1,2j}^{l-1}, v_{2i,2j+1}^{l-1}, v_{2i+1,2j+1}^{l-1} \in \mathcal{V}_{l-1}$.

Attention layer: We exploit the attention mechanism to improve the accuracy. The attention layer consists of *feature attention* that calculates the attention scores for each vertex feature, and *vertex attention* that calculates the attention scores for each vertex. The feature attention is calculated by:

$$\begin{aligned}\boldsymbol{F}_{\text{fa}} = \text{sigmoid}(&\text{mean}(\{\boldsymbol{h}_{i,j} : v_{i,j} \in \mathcal{V}\})\boldsymbol{W}_{\text{fa}}^{\text{mean}} + \\ &\text{sum}(\{\boldsymbol{h}_{i,j} : v_{i,j} \in \mathcal{V}\})\boldsymbol{W}_{\text{fa}}^{\text{sum}})\end{aligned} \quad (2)$$

where $\boldsymbol{h}_{i,j}, \boldsymbol{F}_{\text{fa}} \in \mathbb{R}^{c}$, $\boldsymbol{W}_{\text{fa}}^{\text{mean}}, \boldsymbol{W}_{\text{fa}}^{\text{sum}} \in \mathbb{R}^{c\times c}$, and c denotes the length of feature vector. $\text{fa}[i]$ is the attention score for i^{th} feature. The vertex attention score is calculated using a GNN layer:

$$\{\alpha_{i,j} : v_{i,j} \in \mathcal{V}_l\} = \text{sigmoid}(\text{GNNL}(\{\boldsymbol{h}_{i,j} : v_{i,j} \in \mathcal{V}_{l-1}\})), \quad (3)$$

Where $\alpha_{i,j}$ is the attention score for vertex $v_{i,j}$. Then, the output of the attention layer is calculated by:

$$\{\boldsymbol{h}_{i,j}^{\text{out}} : \boldsymbol{h}_{i,j}^{\text{out}} = (1 + \alpha_{i,j})\boldsymbol{h}_{i,j}^{\text{in}} + \boldsymbol{h}_{i,j}^{\text{in}} \otimes \boldsymbol{F}_{\text{fa}}\} \quad (4)$$

where $\otimes$ is element-wise multiplication.

Multi-layer Perceptron (MLP): After a sequence of layers, all the feature vectors are flattened into a single vector, which is sent to the MLP for classification. MLP has a stack of fully connected (FC) layers.

III. PRUNING

This section covers the proposed pruning techniques, including, input pruning (Section III-A), and weight pruning (Section III-B).

A. Input Pruning

The key benefit of using GNN is that GNN is flexible in accepting any graph structure as the input. Thereby, we are able to exploit input pruning to reduce the computation complexity. Theoretically, in a SAR image (See Figure 1), the pixels not in the target do not affect the classification results. As studied in [26], by properly setting up a constant threshold I_v, we can filter out most irrelevant pixels since the pixels that do not belong to the target usually have negligible SAR signal magnitude. After constructing the input graph from the SAR image, we prune the vertices that have a magnitude smaller than I_v. The magnitude of a vertex is calculated by $\sqrt{x_1^2 + x_2^2 + ... + x_{np}^2}$ where np denotes the number of polarization of the SAR signal. For example, a quad-polarization system has four kinds of polarization – horizontal-horizontal (HH), vertical-vertical (VV), horizontal-vertical (HV), and vertical-horizontal (VH). After pruning the vertices, all the edges connected to the pruned vertices are also pruned. Due to the input pruning, the graph pooling operation (Equation 1) is slightly modified:

$$\begin{aligned}\boldsymbol{h}_{i,j}^{l} = \max(&\mathbb{1}_{2i,2j}^{l-1} \cdot \boldsymbol{h}_{2i,2j}^{l-1}, \mathbb{1}_{2i+1,2j}^{l-1} \cdot \boldsymbol{h}_{2i+1,2j}^{l-1}, \\ &\mathbb{1}_{2i,2j+1}^{l-1} \cdot \boldsymbol{h}_{2i,2j+1}^{l-1}, \mathbb{1}_{2i+1,2j+1}^{l-1} \cdot \boldsymbol{h}_{2i+1,2j+1}^{l-1}),\end{aligned} \quad (5)$$

where $\mathbb{1}_{i,j} \in \{0,1\}$ is the indicator that indicates the existence of vertex $v_{i,j}$. After input pruning, we can skip the computation for the pruned vertices, which greatly reduces the total computation complexity.

B. Weight Pruning

As analyzed in [27], [28], the weight matrices in GNNs have redundancy, and some weight entries can be pruned without affecting the classification accuracy. Therefore, to reduce the total computation complexity, we perform weight pruning by training the model using lasso regression [29]. We add the L1 penalty to the loss function:

$$\text{loss} = l(y, y') + \lambda \sum_{w}^{W} |w| \quad (6)$$

where $l(y, y')$ is the classification loss, and $\lambda \sum_w^W |w|$ is the L1 penalty term parameterized by λ. The L1 penalty leads to weight shrinkage during training. Thereby, some model weights become zeros and can be eliminated from the model. After training, we set a threshold I_w, and the model weights with absolute values smaller than I_w are pruned.

IV. Evaluation

We evaluate our approach on two widely used datasets:

- **MSTAR**: The setting of the MSTAR dataset follows the state-of-the-art work [11] [14] [15] [12]. MSTAR contains the SAR images of ten classes of ground vehicles, with 2747 images in the training set and 2427 images in the testing set.
- **Ship discrimination [30]:** For the ship discrimination dataset, we follow the setting in [31], which is a binary classification task that identifies if a given SAR image has a ship or not. The dataset contains 1596 positive image samples and 1596 negative image samples.

A. Evaluation on MSTAR Dataset

1) Experimental Setting: For the MSTAR dataset, we use the following setting. The proposed model consists of 12 layers. We develop the proposed model using Pytorch Geometric. We use the cross-entropy loss as the classification loss (Equation 6). We train the model using the Adam optimization algorithm. The training batch size is set as 20, and the initial learning rate is 0.02. λ (for lasso regression) is set as 0.002. The L2 weight decay is set as 0.08. We train the model for 150 epochs, and the learning rate is multiplied by 0.5 for every 10 epoch. We use the 8-connectivity to build the input graph. We evaluate the three widely used GNN layers in the proposed model – GCN layer [20], GraphSAGE layer [21], and GAT [22]. We train the proposed model using one NVIDIA RTX A6000 GPU.

Performance metrics: We evaluate the proposed approach using the following metrics: *classification accuracy*, *computation complexity*, and *number of parameters*.

TABLE II. THE ACCURACY ON MSTAR DATASET

GNN Layer Type	Connectivity	Training Accuracy	Testing Accuracy	Training Time
GCN	4	99.16%	90.06%	3.0 hours
	8	95.44%	83.82%	4.0 hours
GAT	4	99.53%	92.21%	1.8 hours
	8	82.71%	71.33%	1.9 hours
GraphSAGE	4	100.00%	97.81%	52 min
	8	100.00%	99.38%	55 min

2) Classification Accuracy: The accuracy of the proposed model (under various GNN layer types and connectivity) is shown in Table II. We observe that using the GraphSAGE layer as the GNN layer leads to the highest training/testing accuracy. Using the GraphSAGE layer also leads to the lowest training time. For the GraphSAGE layer, using 8-connectivity to build the input graph can result in higher accuracy but slightly higher training time than 4-connectivity. Table III shows that the proposed GNN model achieves higher accuracy compared with the state-of-the-art CNNs [11], [12], [14], [15] with negligible computation complexity for inference.

TABLE III. COMPARISON WITH THE STATE-OF-THE-ART CNNS ON MSTAR DATASET

	Type	Accuracy	# of FLOPs	# of Para.
[11]	CNN	92.3%	$\frac{1}{12}\times$	0.5×10^6
[14]	CNN	97.97%	$\frac{1}{10}\times$	0.65×10^6
[15]	CNN	98.52%	$\frac{1}{3}\times$	2.1×10^6
[12]	CNN	99.3%	1× (6.94 GFLOPs)	2.5×10^6
This work [after pruning] (GraphSAGE layer, 8-connectivity)	GNN	99.1%	$\frac{1}{3000}\times$	0.03×10^6

TABLE IV. THE IMPACT OF THE ATTENTION MECHANISM (USING GRAPHSAGE LAYER AND 8-CONNECTIVITY)

Vertex Attention	Feature Attention	Training Accuracy	Testing Accuracy	Training Time
✗	✗	99.67%	93.77%	31 min
✗	✓	100.0%	98.51%	40 min
✓	✗	100.0%	99.26%	41 min
✓	✓	100.0%	99.38%	55 min

3) Ablation Study: We perform an ablation study to evaluate the impact of the attention mechanism (using GraphSAGE layer and 8-connectivity). The result is shown in Table IV. Without vertex and feature attention, the model achieves only 93.77% accuracy. With only vertex attention, the model achieves 99.26% accuracy. With only feature attention, the model achieves 98.51% accuracy. With both vertex and feature attention, the model achieves 99.38% accuracy. The evaluation result demonstrates that the attention mechanism can improve classification accuracy without significantly increasing computation complexity.

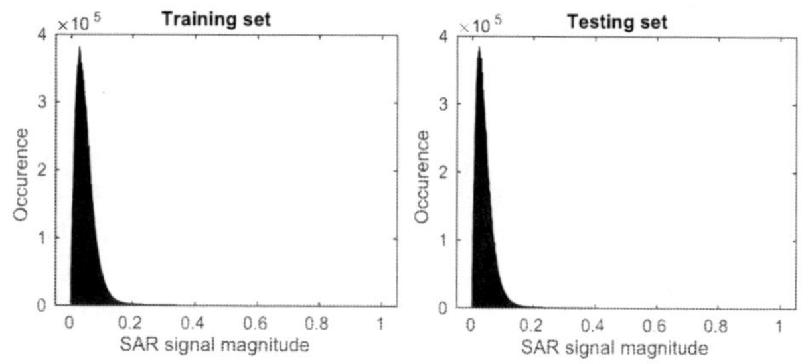

Figure 6. The distribution of the SAR signal magnitude in the training/testing set of MASTAR

4) Evaluation on the Pruning Strategy: We evaluate the proposed input pruning and weight pruning strategies. We use GraphSAGE layer and 8-connectivity as the setting of the model.

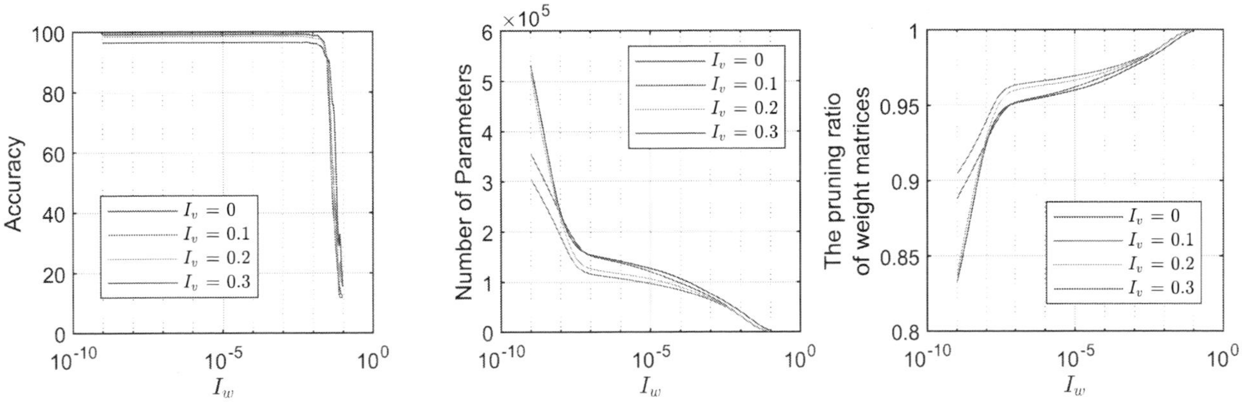

Figure 7. Evaluation of proposed pruning strategy

Input Pruning: Figure 6 shows the data distribution of the SAR signal magnitude of the image pixels in the training/testing set. The SAR signal magnitude ranges from 0 to 16. Since most pixels have a magnitude between 0-1, Figure 6 only shows the range 0-1. For experiment, we set the pruning threshold I_v (See Section III-A) to be 0, 0.1, 0.2, 0.3 respectively. The image pixels that have a magnitude small than I_v are pruned.

Weight Pruning: The weights in weight matrices can be either negative or positive. We set the threshold I_w for weight pruning (See Section III-B). The weights that have an absolute value that is smaller than I_w are pruned. In the experiment, we set I_w to be between 1×10^9 and 1×10^1.

The evaluation results for the pruning strategy are shown in Figure 7. We have the following observations:

- Without weight pruning, when $I_v = 0.1$, 93.4% input vertices/pixels are pruned, the accuracy is dropped to 99.1%; when $I_v = 0.2$, 98.6% input vertices/pixels are pruned, the accuracy is dropped to 98.5%; when $I_v = 0.3$, 99.1% input vertices/pixels are pruned, the accuracy is dropped to 96.5%.
- When weight pruning threshold $I_w < 10^7$, the accuracy does not change w.r.t. to I_w. When $I_w = 10^7$, more than 95% weights are pruned. Therefore, most entries in the weight matrices are redundant.

Therefore, by setting proper threshold I_v, I_w for input pruning and weight pruning, most input pixels and weights can be pruned without significantly dropping the accuracy. Figure 7 shows the evaluation results for the pruning strategy, 97% weight entries are pruned, and the accuracy is 99.1%. By skipping the computation for the pruned vertices and weights, we can dramatically reduce the total computation complexity.

5) Experimental Setting: For ship discrimination dataset, we follow the setting of [31] to conduct experiment for few-shot learning. Since the ship discrimination is a binary class task, the few-shot learning task can be formed as a 2-way-K-shot-classification problem, where $K = \{1, 2, .., 10\}$ denotes the number of labeled training images for each class. We train the model using the Adam optimization algorithm. The training batch size is set as $\frac{K}{2}$, and the learning rate is set as $0.001 * K$. The L2 weight decay is set as 0.08.

6) Classification Accuracy: As shown in Figure 8, we compare our accuracy with [31] (baseline) for the few-shot learning on the ship discrimination dataset. Note that the baseline [31] uses a convolutional neural network (CNN), and the authors pretrained their CNN using the ship discrimination dataset on the Electro-Optical (EO) domain. We do not pretrain our network on any dataset. For various K, the proposed model outperforms the baseline [31], which is a pretrained deep CNN model.

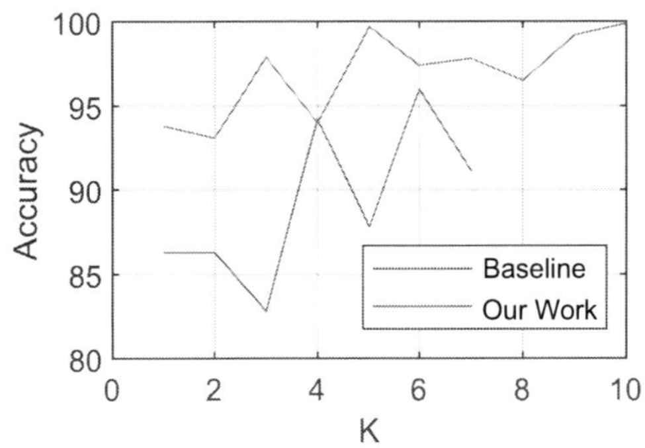

Figure 8. The accuracy on the ship discrimination dataset

TABLE V. COMPARISON OF ACCURACY (%)

K	1	2	3	4	5	6	7
Baseline [31]	86.3	86.3	82.8	94.2	87.8	96.0	91.1
Our work	93.8	93.1	97.9	94.0	99.7	97.4	97.8

V. Conclusion and Future Work

In this paper, we proposed a novel GNN-based approach for SAR automatic target recognition. The proposed approach uses the GNN layer as the backbone and uses the attention mechanism to improve classification accuracy. We proposed pruning strategies, including input pruning and weight pruning, to reduce the computation complexity. The evaluation results on the MSTAR and ship discrimination datasets show that the proposed model outperforms the state-of-the-art CNNs in classification accuracy and computation complexity. In [32], we designed a hardware accelerator for the proposed GNN model. In the future, we plan to extend the proposed GNN model to more SAR-related tasks, such as object detection.

ACKNOWLEDGMENT

This work is supported by the National Science Foundation (NSF) under grants CCF-1919289 and OAC-2209563, and the DEVCOM Army Research Lab (ARL) under grant W911NF2220159.

REFERENCES

[1] L. Landuyt, A. Van Wesemael *et al.*, "Flood mapping based on synthetic aperture radar: An assessment of established approaches," *IEEE Transactions on Geoscience and Remote Sensing*, 2018.

[2] P. Zhan, W. Zhu, and N. Li, "An automated rice mapping method based on flooding signals in synthetic aperture radar time series," *Remote Sensing of Environment*, vol. 252, p. 112112, 2021.

[3] N. Li, Z. Guo *et al.*, "Characterizing ancient channel of the yellow river from spaceborne sar: Case study of chinese gaofen-3 satellite," *IEEE Geoscience and Remote Sensing Letters*, vol. 19, pp. 1–5, 2021.

[4] T. Zhang, X. Zhang *et al.*, "Hyperli-net: A hyper-light deep learning network for high-accurate and high-speed ship detection from synthetic aperture radar imagery," *ISPRS Journal of Photogrammetry and Remote Sensing*, vol. 167, pp. 123–153, 2020.

[5] F. Bardi, W. Frodella *et al.*, "Integration between ground based and satellite sar data in landslide mapping: The san fratello case study," *Geomorphology*, vol. 223, pp. 45–60, 2014.

[6] H. Septanto and O. Sudjana, "Simulation-based energy balance analysis of sar micro-satellite," in *Sixth International Symposium on LAPAN-IPB Satellite*, vol. 11372. International Society for Optics and Photonics, 2019, p. 113721W.

[7] Y. Yokota, Y. Okada *et al.*, "Newly developed x-band sar system onboard japanese small satellite "asnaro-2"," in *Conference Proceedings of 2013 Asia-Pacific Conference on Synthetic Aperture Radar (APSAR)*. IEEE, 2013, pp. 81–83.

[8] P. R. Akbar, H. Saito *et al.*, "Parallel-plate slot array antenna for deployable sar antenna onboard small satellite," *IEEE Transactions on Antennas and Propagation*, 2016.

[9] K. Tanaka, H. Saito *et al.*, "Development of 1kw high power x-band sar installed on small satellite for on-demand observation," in *Proceedings of the International Astronautical Congress, IAC*, 2018.

[10] M. Tsamsakizoglou, H. Löfgren, and M. Gunnarsson, "Microsatellite power control and distribution unit for the innosat platform," in *E3S Web of Conferences*, vol. 16. EDP Sciences, 2017, p. 18007.

[11] M. Zhang, J. An *et al.*, "Convolutional neural network with attention mechanism for sar automatic target recognition," *IEEE Geoscience and Remote Sensing Letters*, 2020.

[12] D. A. Morgan, "Deep convolutional neural networks for atr from sar imagery," in *Algorithms for Synthetic Aperture Radar Imagery XXII*, vol. 9475. SPIE, 2015, pp. 116–128.

[13] J. Hu, L. Shen, and G. Sun, "Squeeze-and-excitation networks," in *Proceedings of the IEEE conference on computer vision and pattern recognition*, 2018, pp. 7132–7141.

[14] J. Pei, Y. Huang *et al.*, "Sar automatic target recognition based on multiview deep learning framework," *IEEE Transactions on Geoscience and Remote Sensing*, 2017.

[15] Z. Ying, C. Xuan *et al.*, "Tai-sarnet: Deep transferred atrous-inception cnn for small samples sar atr," *Sensors*, vol. 20, no. 6, p. 1724, 2020.

[16] H. Zhu, X. Du, and Y. Yao, "Convsppis: identifying protein-protein interaction sites by an ensemble convolutional neural network with feature graph," *Current Bioinformatics*, 2020.

[17] T. Zhao, Y. Hu *et al.*, "Identifying drug–target interactions based on graph convolutional network and deep neural network," *Briefings in bioinformatics*, 2021.

[18] C. R. Qi, L. Yi *et al.*, "Pointnet++: Deep hierarchical feature learning on point sets in a metric space," *Advances in neural information processing systems*, vol. 30, 2017.

[19] K. Xu, W. Hu *et al.*, "How powerful are graph neural networks?" *arXiv preprint arXiv:1810.00826*, 2018.

[20] T. N. Kipf and M. Welling, "Semi-supervised classification with graph convolutional networks," *arXiv preprint arXiv:1609.02907*, 2016.

[21] W. L. Hamilton, R. Ying, and J. Leskovec, "Inductive representation learning on large graphs," in *Proceedings of the 31st International Conference on Neural Information Processing Systems*, 2017.

[22] P. Veličković, G. Cucurull *et al.*, "Graph attention networks," *arXiv preprint arXiv:1710.10903*, 2017.

[23] M. Zhang and Y. Chen, "Link prediction based on graph neural networks," *Advances in Neural Information Processing Systems*, 2018.

[24] R. Ying, J. You *et al.*, "Hierarchical graph representation learning with differentiable pooling," *arXiv preprint arXiv:1806.08804*, 2018.

[25] F. Wu and A. Souza, "Simplifying graph convolutional networks," in *International conference on machine learning*. PMLR, 2019.

[26] H. Zhu, N. Lin *et al.*, "Target classification from sar imagery based on the pixel grayscale decline by graph convolutional neural network," *IEEE Sensors Letters*, vol. 4, no. 6, pp. 1–4, 2020.

[27] M. Rahman, A. Azad *et al.*, "Triple sparsification of graph convolutional networks without sacrificing the accuracy," *arXiv preprint arXiv:2208.03559*, 2022.

[28] H. Zhou, A. Srivastava *et al.*, "Accelerating large scale real-time gnn inference using channel pruning," *arXiv preprint arXiv:2105.04528*, 2021.

[29] R. Tibshirani, "Regression shrinkage and selection via the lasso," *Journal of the Royal Statistical Society: Series B (Methodological)*, vol. 58, no. 1, pp. 267–288, 1996.

[30] C. P. Schwegmann, W. Kleynhans *et al.*, "Very deep learning for ship discrimination in synthetic aperture radar imagery," in *2016 IEEE International Geoscience and Remote Sensing Symposium (IGARSS)*. IEEE, 2016, pp. 104–107.

[31] M. Rostami, S. Kolouri *et al.*, "Deep transfer learning for few-shot sar image classification," *Remote Sensing*, vol. 11, no. 11, p. 1374, 2019.

[32] B. Zhang, R. Kannan *et al.*, "Accurate, low-latency, efficient sar automatic target recognition on fpga," in *2022 32nd International Conference on Field-Programmable Logic and Applications (FPL)*, 2022, pp. 1–8.

Detecting Wildfires Using Unmanned Aerial Vehicle with Near Infrared Optical Imaging Sensor

Edwin Magidimisha
Optronic Sensor systems, Defence and Security,
Council for Scientific and Industrial Research
Pretoria, Country
e-mail: EMagidimisha@csir.co.za

Zimbini Faniso-Mnyaka
Optronic Sensor systems, Defence and Security,
Council for Scientific and Industrial Research
Pretoria, Country
e-mail: ZFaniso@csir.co.za

Seelen Naidoo
Optronic Sensor systems, Defence and Security,
Council for Scientific and Industrial Research
Pretoria, Country
e-mail: SNaidoo@csir.co.za

Muhammad Nana
Optronic Sensor systems, Defence and Security,
Council for Scientific and Industrial Research
Pretoria, Country
mnana@csir.co.za

Abstract— **The increasing severity of wildland fires, largely driven by the effects of climate change, poses a significant risk to both firefighters and ecosystems, while also driving up the cost of combating such fires. Therefore, there is a pressing need for surveillance systems that can detect and track wildfires at an early stage. While traditional methods such as infrared sensors on elevated platforms and surveillance from aircraft have been proven to work in many areas, they have limitations in terms of field of view and cost-effectiveness in covering larger areas. To address these challenges, the Council for Scientific and Industrial Research (CSIR) in South Africa is developing a novel solution for tactical forest firefighting operations. The system takes advantage of the near-infrared (NIR) optical imaging by detecting wildland vegetation fires in the NIR region with cost-effective complementary metal-oxide-semiconductor (CMOS) sensors equipped with ultra-narrow band filters. This approach uses a ratio algorithm on the captured images from two bands to identify pixels with raised intensities that are indicative of a fire. The system is designed to be portable, smaller in size, and can be mounted on an unmanned aerial vehicle to provide real-time support to firefighters. This innovative approach has the potential to significantly improve the speed and accuracy of fire detection in areas where traditional methods are not feasible or effective, such as remote or inaccessible locations.**

Keywords-Potassium: Near-infrared, Unmanned Aerial Vehicle; Wildfire, K-line, CMOS.

I. Introduction

Wildland vegetation fire prevention, detection, monitoring, and suppression are key economic and public safety concerns in many parts of the world [1]. The global incidence and severity of wildfires are expected to rise in response to climate change [2, 3, 4]. These wildfire incidents further exacerbate climate change due to CO2 and black aerosols emission. This serves as a strong motivation for the development of optical surveillance systems that can detect and monitor wildfires. Classical remote sensing of vegetation fires has been through the Planckian emission at the Medium Wave Infrared (3-5 µm, MWIR) and the Long Wave Infrared (LWIR) band of the electromagnetic spectrum [5,6,7]. The Short Wave (1 – 2.5µm, SWIR) infrared band was exploited and deployed on AVIRIS platform [8]. However, the IR band can be mostly affected by other heat emitting sources which could act as false alarms [9].

With the advancement of imaging sensors and filter technologies, there is now increased availability of reliable optical commercial-off-the-shelf (COTS) products at affordable cost. The new sensor technologies such as high-resolution CCD and CMOS sensors, provide an opportunity to enhance wildfire detection, monitoring, and reporting. This study presents a compact and cost-effective method of detecting wildland fires using near-infrared (NIR) spectral line emissions from electronically excited Potassium (K) atoms at 766.5 and 769.9 nm, during biomass burning [10,16, 18, 19]. The Potassium spectral lines can be discriminated against any other background by detector systems that are less costly than the longer wavelength, actively cooled instruments more typically used in EO-based active fire studies [11]. Similarly, new airborne sensor platforms, particularly small, unmanned aircraft, or drones, are enabling new applications for airborne fire sensing [12]. The current study integrates the NIR optical payload onto and operates it from an unmanned aerial vehicle (UAV) using remote sensing techniques.

II. Background

In recent years, we have seen great progress in the use of unmanned aerial vehicles with advanced software for forest fire monitoring, detection, and firefighting. The integration of UAVs with remote sensing techniques is aimed at providing rapid, mobile, and low-cost powerful solutions for various fire tasks [12]. Direct forest fire detection by firefighting agencies, as opposed to reports from the public, is typically performed by some combination of fixed detection platforms such as watch towers, aerial detection patrols, and satellite imagery [11]. However, highly elevated platforms are not well suited in area coverage and can result into some areas developing fires unnoticed. Although aircrafts are considered efficient in firefighting, they are expensive to keep airborne for constant monitoring. Compared to fixed ground-based wildfire detection systems UAVs can provide a broader and more accurate perception of the fire from above especially in areas that are inaccessible or considered too dangerous for operations by firefighting crews. During firefighting, UAVs provide eyes from above and can provide important information on the fire progression.

A vision-based UAV mounted system for detecting forest fires that uses both motion and chroma characteristics of fire in the decision rules to improve the reliability and accuracy of fire detection was proposed [13]. Sudhakar et al. [14] proposed a method for forest fire detection through UAVs equipped with an optical and an infrared camera using a LAB color model and a motion-based algorithm followed by a maximally stable extremal regions (MSERs) extraction module. For improved presentation, the extracted forest fire detections were joined with landscape information and meteorological data. Chen et al. [15] used optical and infrared sensors data to train a CNN first for smoke detection and then for fire flame detection.

III. DETECTION PRINCIPLE

Figure 1 below, illustrates a comprehensive outline of the fire detection system. The system incorporates a dual camera to capture and record images of burning biomass fire, specifically vegetation fires containing the Potassium element signature. The captured images are processed using the OSS developed algorithm that is applied during the image processing stage to analyze the pair of images and establish whether a fire has been detected.

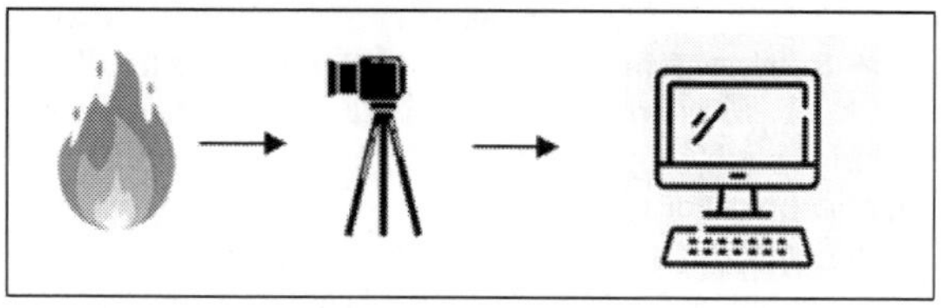

Figure 1: Simplified schematic depicting overview of fire detection system.

A. *The Potassium element*

Potassium belongs to the alkali metal group and is in the first column of the periodic table. It is one of the abundant elements in vegetation species [16, 17]. It has a single valence electron that present unique narrowband spectral emission lines within the visible and near infrared (NIR) wavelength range when vegetation biomass is heated to high temperatures during the process of flaming combustion [18]. The spectral emission of K appears as doublet at 766.5 nm and 769.9 nm spectral bands [19]. With the advancements in filter design, filters can now detect low-level signals while suppressing almost all emissions within the outer band by targeting specific elemental emissions from a source signature. These advancements in technology open the opportunity for the development of compact sensors capable of detecting narrow spectral lines and can be advanced to compete with other passive sensors operating in other bands. In this project, ultra-narrow band imaging is used for the detection of K using CMOS detectors. The integration of COTS, and ultra-narrow band imaging allows the design of compact and less power-hungry systems which can be easily integrated on a UAV.

B. *Fire detection system*

The NIR fire detection sensor presented in this paper comprise of two optical imaging systems placed side-by-side with common (identical) field of view. These cameras are fitted with ultra-narrow band filters with 1nm bandwidth sensitive at 769.9 nm referred to as the K-line band, and 757 nm referred to as the reference band. The target and reference channels are temporally synchronised at the electronic level such that pairs of images (one from the target band and the other from reference channel) are obtained at the same instant. Fires are detected by comparing the NIR channel image to the reference channel image. Pixels which are much brighter in the target channel relative to the reference channel are candidate fire detections.

C. *Image Processing Algorithm*

The system image processing begins when two images are captured, one image with K-emission and the other with the background. The images from the two sensors are captured simultaneously. The reference image is resampled to be aligned with the K-line image pixelwise. This is done by feature mapping, using a Lucas-Kanada optical flow algorithm [20]. The sections of the individual images that are not common in both are then cropped out, leaving two images of the exact same scene.

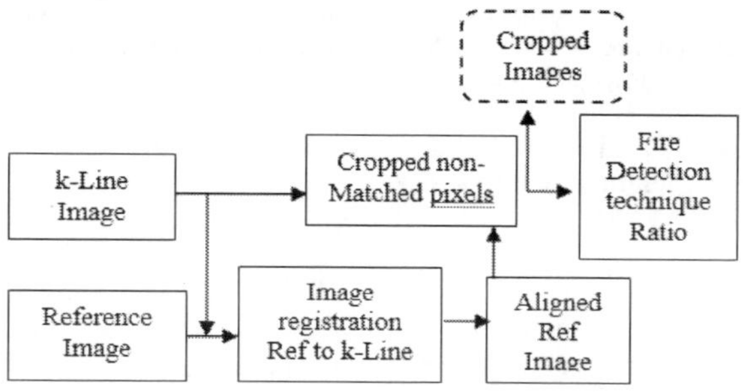

Figure 2: Overview of K-line fire detection principle

The images are not modified with any image enhancement algorithms and are not compressed to preserve the fire front K-line signal emissions. The fire detection algorithm is applied to the matched cropped K-line and reference images. The fire detection is done using the image ratio technique [16, 21]. A block diagram giving an overview of the algorithm is illustrated in Figure 2. The image ratio technique is simple and is implemented as follows:

a) Compute the ratio image i.e., the K-line image divided by the reference image.
b) Compute the global mean and variance of the ratio image.
c) Compute the pixel variance for each pixel in the ratio image.
d) If the variance of a pixel is greater than the global variance multiplied by a user defined sensitivity integer value, the pixel gets classified as a fire pixel. Otherwise, the pixel is classified as a non-fire pixel.

IV. METHOD

The field measurements test was conducted at the Grasslands Flying Club in Pretoria West on the 18th of March 2022. The purpose of the test was to evaluate the aerial performance of the NIR fire detection sensor onboard the UAV. Figure 3 displays a photograph of the NIR imaging sensor system during its lab testing phase.

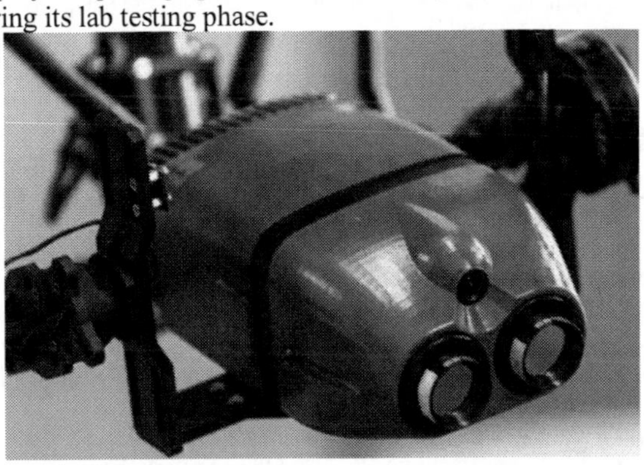

Figure 3: A closer look at the NIR sensor with two CMOS optical sensors placed side by side and furnished with ultra-narrow filters.

The UAV Payload used a development (Raspberry Pi4 8GB) board to control the capturing of images, communication with a ground station, and storage of captured images. The captured images were stored on-board a micro-SD card and removed after the completion of a sortie. When the memory stick is retrieved and the data is retrieved for archival, the data was also inspected while the next mission was ongoing. Fire detection is performed on a post processing basis by automatically analysing the images stored in the memory card. The basic NIR sensor payload consists of the following components:

- a processor module with storage,
- the K-line camera payload,
- a viewfinder camera,
- a telemetry radio downlink,
- an analogue video downlink,
- high-definition video downlink,
- a power source, and
- wiring harnesses.

Figure 4 shows the NIR fire detection payload onboard the UAV taken during the deployment experiment at the Grasslands Flying Club.

Figure 4: UAV with NIR sensor payload on the DJI 600 drone during a field testing of the sensor.

Several sorties where carried A deployment to test the new NIR payload aboard an airborne UAV The purpose of the test was to determine whether the new NIR sensor can detect ground wildfires from the air at relatively low altitudes (approximately 150m above ground level) and at different aspect angles from the fire. The size of the fire on the ground was approximately 500cm by 500cm.

The following equipment was used during the test:

- M600 UAV with RONIN gimbal provided and piloted by UAV Industries (UAVI),
- UAV NIR payload sensor
- UAV Ground Control Station
- FieldSpec 3 Max Analytical Spectral Device (ASD) with spectral range 350-2500nm
- Weather Station

A. *Atmospheric conditions*

During field measurements, the scenario demands that atmospheric computations be made to accommodate the atmospheric effects, caused by molecular absorption and emission (mainly water and carbon dioxide, as well as atmospheric scattering processes by aerosols). The atmospheric modelling codes such as MODTRAN, HITRAN and others can be used to simulate the atmospheric transmission as described below.

The atmospheric transmission was calculated using the HITRAN Radiation Transfer Model (RTM) in the NIR region, as shown in Figure 5. The HITRAN data downloaded were in vacuum scale and converted to air using the Elden equation (NIST). The following parameters were used: 20°C air temperature, 101325pa air pressure and 50% humidity percentage [17]. The red lines show the Potassium doublet at 766.48nm and 769.89nm. The 766.48 is absorbed by atmospheric Oxygen as it is located at the Oxygen absorption line and therefore cannot be detected. The K emission lines are within the range of the sequence of the atmospheric absorption lines that peak at about762 nm [22].

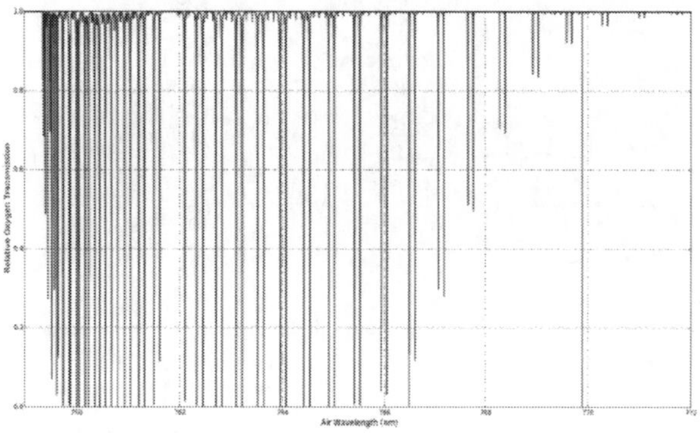

Figure 5: The high spectral resolution Oxygen atmospheric transmittance near the wavelength location of the two Potassium emission spectral lines, data from HiTRAN (http://hitran.iao.ru).

The positions of the K-lines are indicated by two vertical red lines and the deep lines shows the absorption effects arising from the atmospheric absorption gases.

B. *Field UAV measurement*

The test consisted of a controlled ground fire using wood and dried grass as the fuel. An analytical spectral device was setup on the ground close to the fire (approximately 3m) which is used to record the spectral signature of the fire as it burns. It provides reference spectral data of the fire from the ground for checking whether the NIR signature is contained within the fire. The range at which the detection tests are conducted is approximately 150m from the fire. Two test points of interest applicable and sufficient for proving the initial success of the fire detection system is highlighted in the next section. Figure 6 below gives an illustrative overview of the mission profile used during the fire detection tests.

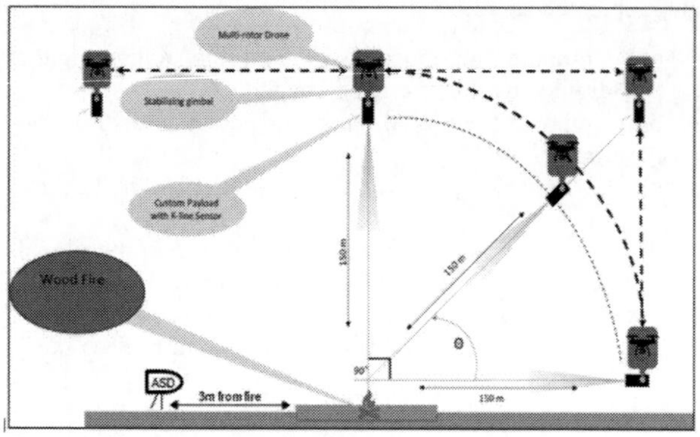

Figure 6: Illustrative overview of flight mission profiles

V. RESULTS

Some of the results from the fire detection tests are presented here. These include UAV NIR image sensor data as well as spectral data recorded by the ASD on the ground. Whenever the NIR image sensor results are shown, they are shown in pairs. The left image is a masked image and the right-hand side is the target image which has the K-line emission signature. The K-line signature is indicated as a red overlay after image processing has been applied. The black and white target image shows the masking and therefore isolating the K-line signature.

The ASD FieldSpec 3 spectroradiometer data is also presented as a pair of images with the left hand side image showing the total spectral image of the fire in the 350-2500 nm spectral band. The right-hand side shows the zoomed unresolved K-line doublet. The various colour lines shown in the total spectral graph indicate the different spectral graphs of the fire at different instances as it burns within a short timeframe (such as during a testpoint recording). The emission spectrum of the fire is clearly visible with spectral radiance generally increasing with wavelength.

Various flight profiles were flown as shared below.

A. *Test Point 1:45 Degree Aspect Angle Fire Detection*

The sensor was able to detect fire from an angle (in this scenario the angle 45° is used). The images were captured while the drone was at 45° as shown in Figure 6. Figure 7 (a) is the masked image of the K-line emission from the fire while 7 (b) shows the unmasked image with K-line painted in red.

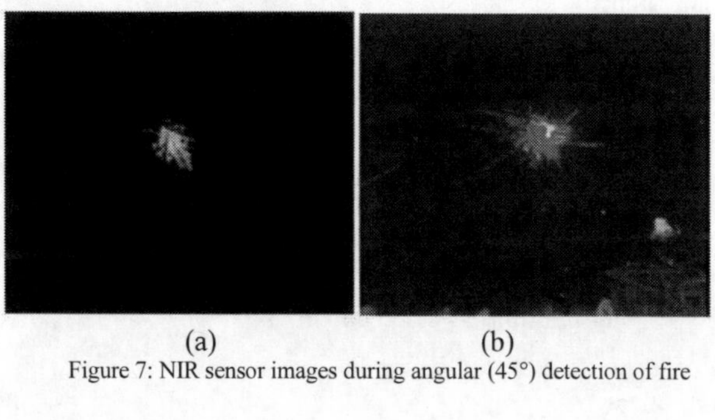

(a) (b)

Figure 7: NIR sensor images during angular (45°) detection of fire

The NIR signature was fully detected by the ASD spectral sensor as it was placed close to the scene of the fire. The ASD data is given in Figure 8 below. Multiple ASD spectral measurements were taken while the UAV was airborne, as shown in Figure 8(a) and 8(b).

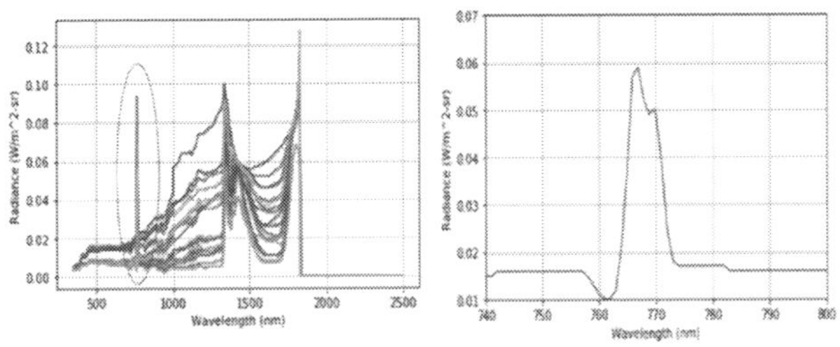

(a) (b)

Figure 8: ASD spectral data with NIR zoomed K-line unresolved doublet. The figure shows the spectral radiance of the fire within the NIR region.

Figure 8(b) is the zoomed spectra showing the unresolved K-line doublet due to the low ASD resolution of 3nm.

B. Test Point 2: Flying directly Above the Fire (90° aspect angle)

The sensor was able to detect fire from directly above as shown in Figure 9. The figure shows the NIR images as detected by the K-line band.

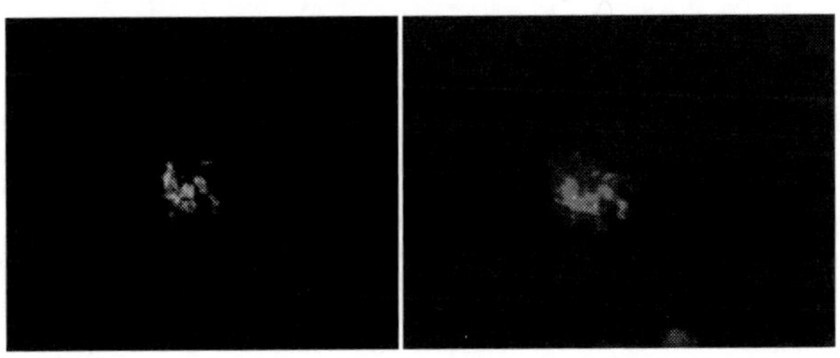

Figure 9: NIR sensor images showing fire detection from directly above

Data logging was a success on the ASD sensor however, the results show that the K-line signature was not strong (figure 10). This is suspected to be because the fire was not burning strong after adding more wood.

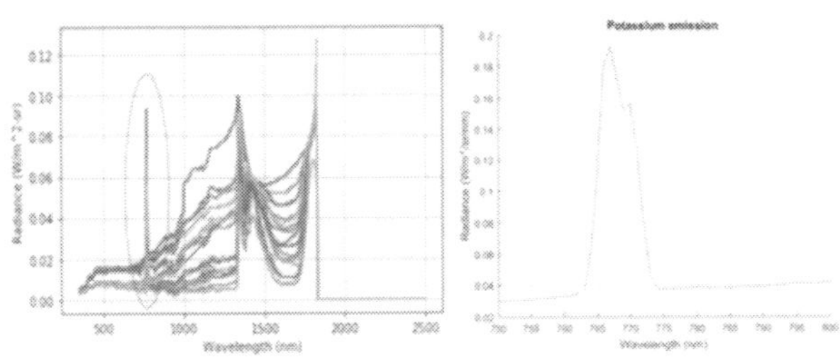

(a) (b)

Figure 10: ASD data with NIR doublet extracted (b) unresolved K-line doublet.

VI. Conclusion

Small scale fires were captured using a K-line based fire detection sensor mounted on an unmanned aerial vehicle during a field trial at the Centurion Flying club, Pretoria, South Africa. Results present strong evidence of K-line signature within the vegetation fires detectable by compact CMOS cameras operating within the NIR. The ASD measurement confirmed the spectral location of the K alkali metal present on the vegetation biomass.

This study provides the possibility to perform early fire detection of vegetation biomass using low cost NIR sensors integrated on unmanned aerial vehicles coupled with advanced image processing algorithms. This work is recommended as work in progress to develop system that will not only detect but geo-locate, monitor fire progression, and make evolution estimates of fires in real-time.

Acknowledgment

The Optronic Sensor Systems (OSS) together with all the team members involved in this work, owes thanks to the Aeronautical Sciences Impact Area at the CSIR for the technical assistance provided and to the Department of Science and Innovation (DSI) for their financial support and funding for enabling this research effort.

References

[1] Robert S. A., M. Joshua, J. G. Craig, and S. Jennings, Airborne Optical and Thermal Remote Sensing for Wildfire Detection and Monitoring, 2016.

[2] Liu, Y., J.A. Stanturf, and S.L. Goodrick. Trends in global wildfire potential in a changing climate. Forest Ecology and Management, 259(2010):685–697, 2009. http://dx.doi.org/10.1016/j.foreco.2009.09.002.

[3] Kelly, R., L. Melissa. C, Philip, E. Higuera, I. Stefanova, B. L. Brubaker, and F. Sheng Hu. Recent burning of boreal forests exceeds fire regime limits of the past 10,000 years. Proceedings of the National Academy of Sciences, 110(32):13055–13060, 2013. http://dx.doi.org/10.1073/pnas. 1305069110.

[4] Philip E. Dennison, Dar A. Roberts, and Lucy Kammer. Wildfire Detection for Retrieving Fire Temperature from Hyperspectral Data. In ASPRS 2008 Annual Conference, volume 1, pages 139–146, 2008. http: //www.asprs.org/a/publications/proceedings/portland08/0015.pdf

[5] Robinson, J.M., Fire from space: global fire evaluation using infrared remote sensing. International Journal of Remote Sensing, 12, 3-24, 1991.

[6] Fuller, D.O., Satellite remote sensing of biomass burning using optical and thermal sensors, Progress in Physical Geography, 24, 543-561, 2000.

[7] Lentile, L.B., Z.A. Holden, A.M. Smith, M.J. Falkowski, A.T. Hudak, P. Morgan, S.A. Lewis, P.E. Gessler, and N.C. Benson, Remote sensing techniques to assess active fire characteristics and post fire effects, International Journal of Wildland Fire, 15, 319-345, 2006.

[8] Thomas, P.J., and O.N, Near-infrared forest fire detection concept, Applied Optics, 32, 5348-5355, 1993.

[9] Wang, Z., Modelling Wildland Fire Radiance in Synthetic Remote Sensing Scenes, PhD thesis, 2007.

[10] Magidimisha, E., and D, Griffiths, Remote optical observations of actively burning biomass fires using potassium line emission, Proceedings of the SPIE, Volume 10036, id. 1003611 6 pp, 2016.

[11] Stefania A, J. Martin, b. Wooster, and A. Piscini, Multi-resolution spectral analysis of wildfire potassium emission signatures using laboratory, airborne and spaceborne remote sensing, Remote Sensing of Environment, 115, 1811–1823, 2011.

[12] Allison, R.S., A, J. M. Johnston, G. Craig, and S. Jennings, Airborne Optical and Thermal Remote Sensing for Wildfire Detection and Monitoring, Sensors 2016.

[13] Yuan, C., Z. Liu and Y. Zhang, Vision-based Forest Fire Detection in Aerial Images for Firefighting Using UAVs, 2015.

[14] Sudhakar, S. V. Vijayakumar, C.S Kumar, V. Priya, L. Ravi, V. Subramaniya swamy, Unmanned Aerial Vehicle (UAV) based Forest Fire Detection and monitoring for reducing false alarms in forest-fires. Comput. Commun, 149, 1–16, 2020.

[15] Chen, Y., Y. Zhang, J. Xin, Y. Yi, D. Liu, H. Liu, A UAV-based Forest Fire Detection Algorithm Using Convolutional Neural Network. In Proceedings of the IEEE 37th Chinese Control Conference,Wuhan, China, 25–27 July, pp. 10305–10310, 2018.

[16] Vodacek, A., Kremens, R. L., Fordham, A. J., Vangorden, S. C., Luisi, D., Shott, J. R., &Latham, D. J. Remote optical detection of biomass burning using a potassium emission signature. International Journal of Remote Sensing, 23(3), 721–2726, 2002.

[17] Nist Atomic Spectral Database. URL.http://Physics.nist.gov, 2001.

[18] S.Amici, M.J. Wooster, and A. Piscini. Multi-resolution spectral analysis of wildfire potassium emission signatures using laboratory, airborne and spaceborne remote sensing, 2011.

[19] Latham, D., 1998, Near-infrared spectral lines in natural !res. Proceedings of the III International Conference on Forest Fire Research/14th Conference on Fire and Forest Meteorology, 16–20 November 1998 (Coimbra, Portugal: ADAI), pp. 513–515.

[20] "*OpenCV Tutorial Optical Flow*" docs.opencv.org. https://docs.opencv.org/4.5.1/d4/dee/tutorial_optical_flow.html (accessed Apr. 4, 2023).

[21] Ononye, A. E., A. Vodacek, and R. Kremens, Fire temperature retrieval using constrained spectral unmixing and emissivity estimation, Algorithms and Technologies for Multispectral, Hyperspectral, and Ultraspectral Imagery XI **5806**, 352 – 360, [doi: 10.1117/12.603440], 2005.

[22] Pearse, R.W.B., and A.G. Gaydon, The identification of Molecular Spectra (London: Chapman and Hall), 1976.

Dynamic Integration of Climate Properties with Geospatial Data for Energy Applications

Lassi Lehto and Jaakko Kähkönen
Department of Geoinformatics and Cartography
Finnish Geospatial Research Institute
Espoo, Finland
e-mail: lassi.lehto@nls.fi, jaakko.kahkonen@nls.fi

Abstract— **The current energy crisis affects the whole Europe, making cross-border provision of climate data indispensable. Energy-related applications are thus becoming increasingly significant also for geospatial service development. Climate properties of the study areas are often critical for informed decisions in energy applications. Energy and climate are selected as important use cases in the GeoE3 (Geospatially Enabled Ecosystem for Europe) project involving five European countries. The cloud-based data integrations platform developed in the project facilitates harmonized access to climate properties provided by meteorological institutes in all of the participating countries. Various download and direct access methods are applied in getting the climate data to the platform, from which it is offered to the client applications via the OGC API Coverages service interface.**

Keywords-climate data; cross-border; harmonization; OGC API Coverages.

I. Introduction

Consistent cross-border delivery of harmonized climate data has become indispensable with the current energy crisis ravaging whole Europe. It is particularly important to provide access to climate properties in connection of geospatial services [1][2], for example to support energy-related decisions in spatial planning and in the construction industry. Buildings are an important factor in energy-saving endeavors and it is thus critical to provide climate properties readily connected with building data.

The Geospatially Enabled Ecosystem for Europe (GeoE3) project has been developing a cloud-based service platform for cross-border and cross-domain integration of geospatial data resources [3]. Climate data has been selected from the beginning as one of the key data sources in the project. Renewable energy is one of the three use cases of GeoE3, making buildings a central content theme for the project.

GeoE3 involves five European countries: Finland, Estonia, Norway, the Netherlands and Spain. National Mapping and Cadastral Agencies (NMCAs) of each country are participating as members of the project consortium. Cross-border access to 3D building models has been one of the main goals of the project. This goal has been quite successfully achieved, as 3D buildings from all members countries are now available from the GeoE3 integration platform [4].

The national meteorological institute is present in the project consortium only in case of Finland. However, open Application Programming Interfaces (APIs) have been made available by the Norwegian, Dutch and Spanish national meteorological institutes. Dedicated access modules have been developed for the integration platform to accommodate the varying meteorological APIs available.

Climate data is made available to client applications from the platform via the modern Open Geospatial Consortium's (OGC) OGC API Coverages access interface that also enables climate properties to be attached to individual buildings [5].

The rest of the paper is organised as follows. Section II introduces the GeoE3 integration platform. Section III discusses the solutions used for accessing climate data from the five participating countries. Section IV describes the approach applied for the harmonized provision of climate data via the OGC API Coverages service interface. The paper ends with conclusions in Section V.

II. Integration Platform

The GeoE3 integration platform is a cloud-based service providing single point access to all thematic content offered by the project [6]. The platform is essentially based on pygeoapi, a Python implementation of the OGC API family of service interface standards [7]. On the GeoE3 platform, pygeoapi has been adapted to run on top of the web service development framework called Django [8]. In addition to the selected climate-related content themes: Temperature, Windspeed and Sunshine, the platform also provides harmonized access to the following themes: 2D/3D buildings, 2D/3D roads, Digital Terrain Model (DTM) and Digital Surface Model (DSM). At the moment the platform contains altogether 42 datasets/services (2D buildings: 7, 3D buildings: 7, 2D roads 4, 3D roads: 2, DTM: 6, DSM: 5, temperature: 5, windspeed: 4, sunshine: 2).

Dedicated content provider modules have been developed to access national legacy service end points providing data in national schemas. The integration platform takes care of all necessary harmonization procedures to make the content offering consistent across participating countries, and makes harmonized content available to client applications via the service interfaces conforming to the modern OGC API family of access standards. Physically the integration platform runs on the cloud service environment cPouta of the Finnish IT Center for Science (CSC) in Kajaani, Finland [9].

III. Access Methods

The development for the provision of climate data service interface on the GeoE3 integration platform started with the Finnish data, as Finnish Meteorological Institute (FMI) is one of the project members [10]. Access to meteorological APIs in Norway and the Netherlands was found via the respective NMCAs. With the help of the Spanish NMCA, an API facilitating download of climate data sets from Spain was then found. A data repository maintained by the FinEst Centre for Smart Cities finally enabled provisional access also to Estonian climate data resources.

The solution for climate data access provided by FMI is based on OGC API Features implementation. The climate properties available from the FMI API include 30 years (1991-2020) averages of temperature, windspeed and yearly sunshine hours [11]. The resulting data set includes temperature values from 99 observation stations, windspeed values from 25 stations and sunshine hours from 8 stations.

From the station values and positions an interpolated grid representation is computed using Python libraries Pandas [12] and SciPy [13]. The interpolation method 'linear' is used, where observation station positions makes it possible, and the 'nearest' operation is applied for the rest of the country's area. The result is saved as a GeoTIFF image file on the integration platform. The stored image is used as the source of the platform's content provision via OGC API Coverages until the file is removed, when the FMI API is accessed again. The same base processing pipeline (see Figure 4) is followed with all the climate themes across all countries' meteorological APIs.

In the case of Norway, a custom API called Frost was used, provided by MET Norway [14]. The API requires registration and use of an API key in all data access requests. The API follows the principles established for a RESTful API and provides the results in JSON encoding. The API provides access to 30 years temperature values from 275 observation stations and windspeed values from 230 stations. Long period averages are computed by the integration platform for the period 1991-2020 and the grid representation then saved into the GeoTIFF file.

Royal Netherlands Meteorological Institute has made available a data access solution called KNMI Data Platform [15]. It offers a flexible API-based file download mechanism. At the moment temperature is the only climate property accessed from KNMI Data Platform. The value is available for 28 observation stations and represents 30 years averages (1981-2010). Values are retrieved from the downloaded NetCDF file using a Python library netCDF4, interpolated into a grid representation and stored as a GeoTIFF file to the integration platform.

The Spanish national meteorological institute Agencia Estatal de Meteorología (AEMET) provides an open access data platform AEMET OpenData [16]. After registration, an API key is provided for accessing the site. The platform offers an Open API -based access interface for retrieval of 30 years' (1981-2010) average temperature values from 105 observation stations, windspeed values from 103 stations and sunshine values from 85 stations. The values are accessed via the API by station, yearly averages are computed from monthly values, and various unit conversions are applied. The climate values are finally interpolated and saved to the GeoTIFF file.

In the case of Estonia, the service development is still ongoing. Historical weather observations are available from the FinEst Centre for Smart Cities [17]. Observation stations datasets are manually downloaded as text files. The files are then processed to retrieve temperature averages for 98 observation stations and windspeed averages for 39 stations. At the moment the Estonian dataset only covers a short period of a few months in 2022.

IV. Service Provision

On the GeoE3 integration platform, Climate data is offered to client applications via the OGC API Coverages service interface. This enables clients to retrieve grid representations for various image-based applications and analysis processes. The platform offers a basic visualization functionality for climate themes, together with simple download options (see Figure 1).

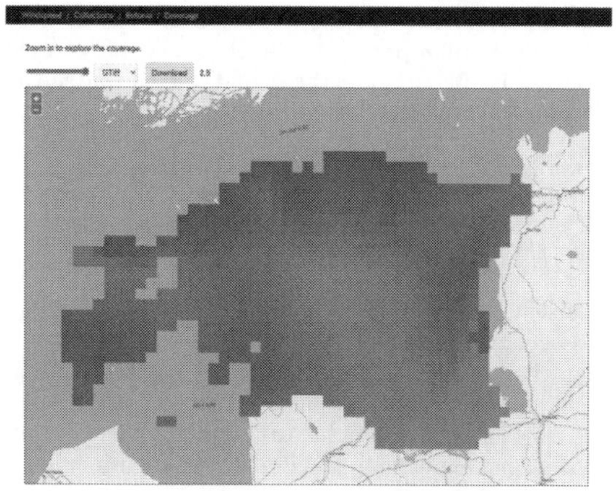

Figure 1. Climate coverage Windspeed of Estonia, visualized in the user interface of the GeoE3 integration platform.

Participating countries are treated as individual data collections inside the single theme-specific OGC API Coverages service instance. For example, the theme 'Temperature' contains currently five data collections, one for each participating country. This arrangement enables natural cross-border data integration inside a single service end point. The current availability of climate properties is shown in Table 1.

TABLE I. AVAILABILITY OF CLIMATE PROPERTIES.

	Temperature	Windspeed	Sunshine
Finland	Yes	Yes	Yes
Estonia	Yes	Yes	
Norway	Yes	Yes	
The Netherlands	Yes		
Spain	Yes	Yes	Yes

Via the OGC API Coverages interface, climate properties can also be queried at arbitrary point locations across the whole data extent. This enables these properties to be connected as additional attributes for example to individual building features. On the GeoE3 integration platform, this connection is made to the 2D buildings. An example of a Finnish building with climate attributes displayed is presented in Figure 2.

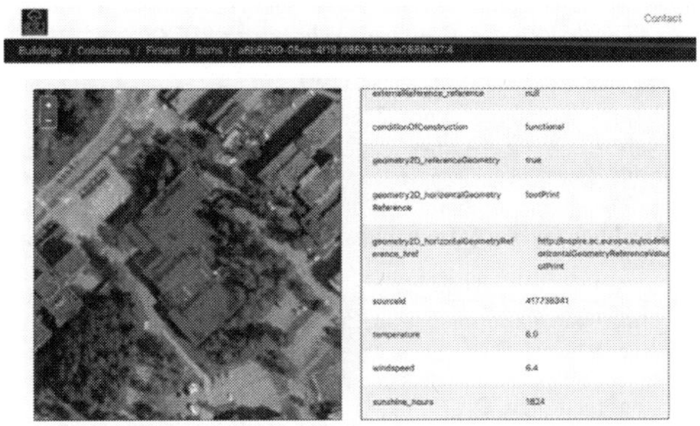

Figure 2. The selected Finnish 2D building shown with the climate-related attributes.

Map visualization of the current integration platform's user interface is based on OpenLayers library [18]. Support for OGC API Coverages has been added by custom code making use of OpenLayers source class ImageStatic.

Figure 3. Integrated temperature visualization covering the area of Finland and Norway.

The so called cross-collection query mechanism facilitates cross-border content queries to be made. Currently this mechanism is officially defined only for the OGC API Features interface specification [19]. In the GeoE3 project, an experimental cross-collection query support has been developed for OGC API Coverages service. The query can be formulated as follows:

```
.../geoe3/temperature/search?
  f=png&
  collections=temperature_NO,temperature_FI&
  subset=x(396919:3888202),y(7701052:11580281)&
  scale-size=x(900),y(1000)
```

With the 'collections' parameter, the desired data collections can be identified. If the query extent, defined by the parameter 'subset', crosses a country boundary, the resulting image will be an integrated visualization of the output of the involved countries. Consistent color maps have been applied for all countries' climate themes, making the visualizations comparable across national borders. As an example, the image resulting from the above query is shown in Figure 3.

V. Conclusion

A dynamic content integration mechanism has been established on the GeoE3 integration platform. Participating countries' datasets are treated as individual data collections inside an OGC API Features or OGC API Coverages service end point. This arrangement enables natural cross-border integration of geospatial content. This kind of integrated dataset can be accessed from the platform employing the so called cross-collection query mechanism, officially defined for OGC API Features and experimentally implemented for OGC API Coverages on the GeoE3 integration platform.

Cross-domain integration has been achieved by accessing climate properties representing long period averages from various custom APIs provided by meteorological institutes. A dedicated provider module has been developed on the integration platform for each API to accommodate the differences in service protocol and content encoding. The result is stored in a GeoTIFF image as cache on the platform. Harmonized climate properties are then delivered via the OGC API Coverages interface, enabling both raster image and point value retrieval. Point values are integrated by the platform as additional attributes to the building features also available from the platform.

The cross-border and cross-domain content integration functionality offered by the GeoE3 integration platform facilitates applications requiring access to harmonized, geospatially referenced climate data across Europe. In particular, combined with building datasets, the climate properties support applications aimed at saving of energy or finding new renewable energy solutions.

Acknowledgment

The GeoE3 project is co-financed by the Connecting Europe Facility (CEF) of the European Union with the grant agreement number INEA/CEF/ICT/A2019/2063390.

The authors wish to acknowledge CSC – IT Center for Science, Finland, for computational resources.

REFERENCES

[1] G. Giuliani, S. Nativi, A. Obregon, M. Beniston, and A. Lehmann, "Spatially enabling the Global Framework for Climate Services: Reviewing geospatial solutions to efficiently share and integrate climate data & information", *Climate Services*, Vol. 8, 2017, pp. 44-58, ISSN 2405-8807, https://doi.org/10.1016/j.cliser.2017.08.003.

[2] A. Sharma, S. M. A. Zaidi, V. Chandola, M. R. Allen, and B. L. Bhaduri, "WebGlobe - A cloud-based geospatial analysis framework for interacting with climate data", *BigSpatial 2018: Proceedings of the 7th ACM SIGSPATIAL International Workshop on Analytics for Big Geospatial Data*, Nov 2018, pp. 42–46, https://doi.org/10.1145/3282834.3282835

[3] GeoE3, Geospatially Enabled Ecosystem for Europe, GeoE3 Home Page. [Online]. Available from: https://geoe3.eu, 2023, [retrieved: Mar, 2023]

[4] L. Lehto and J. Kähkönen, "Cross-border Delivery and Web-based Visualization of 3D Buildings", *GEOProcessing 2022, the Fourteenth International Conference on Advanced Geographic Information Systems, Applications, and Services*, Jun 26-30, 2022, Porto, Portugal, pp. 36-40. ISBN: 978-1-61208-983-6

[5] OGC, OGC API – Coverages Home Page. [Online]. Available from: https://ogcapi.ogc.org/coverages/, 2023, [retrieved: Mar, 2023]

[6] L. Lehto, J. Kähkönen, J. Reini, T. Aarnio, and R. Tervo, "Cross-border and Cross-domain Integration of 3D Content in a European Geospatially Enabled Ecosystem", *International Journal on Advances in Intelligent Systems*, ISSN 1942-2679, Vol. 15, no. 1 & 2, year 2022, pp. 1-11, http://www.iariajournals.org/intelligent_systems/

[7] T. Kralidis, pygeoapi Home Page. [Online]. Available from: https://pygeoapi.io, 2023, [retrieved: Mar, 2023]

[8] Django, Django Home Page. [Online]. Available from https://www.djangoproject.com, 2023, [retrieved: Mar, 2023]

[9] CSC, IT Center for Science Ltd. Home Page. [Online]. Available from: https://www.csc.fi, 2023, [retrieved: Mar, 2023]

[10] FMI, Finnish Meteorological Institute Home Page. [Online]. Available from: https://en.ilmatieteenlaitos.fi, 2023, [retrieved: Mar, 2023]

[11] FMI, FMI Climate Data API. [Online]. Available from: http://stat.obs.weather.fmibeta.com/sofp/collections, 2023, [retrieved: Mar, 2023]

[12] Pandas, Pandas Home Page. [Online]. Available from https://pandas.pydata.org, 2023, [retrieved: Mar, 2023]

[13] SciPy, SkiPy Home Page. [Online]. Available from https://scipy.org, 2023, [retrieved: Mar, 2023]

[14] MET Norway, Frost – MET Norway's Climate Data API. [Online]. Available from: https://frost.met.no, 2023, [retrieved: Mar, 2023]

[15] KNMI, KNMI Data Platform. [Online]. Available from: https://dataplatform.knmi.nl, 2023, [retrieved: Mar, 2023]

[16] AEMET, AEMET Open Data. [Online]. Available from: https://opendata.aemet.es/centrodedescargas/inicio, 2023, [retrieved: Mar, 2023]

[17] FinEst, FinEst Centre for Smart Cities. [Online]. Available from: https://www.finestcentre.eu, 2023, [retrieved: Mar, 2023]

[18] OpenLayers, OpenLayers Home Page. [Online]. Available from: https://openlayers.org, 2023, [retrieved: Mar, 2023]

[19] OGC, "OGC API Features – Part 3: Filtering and the Common Query Language (CQL)". [Online]. Available from https://portal.ogc.org/files/96288, 2020, [retrieved: Mar, 2023]

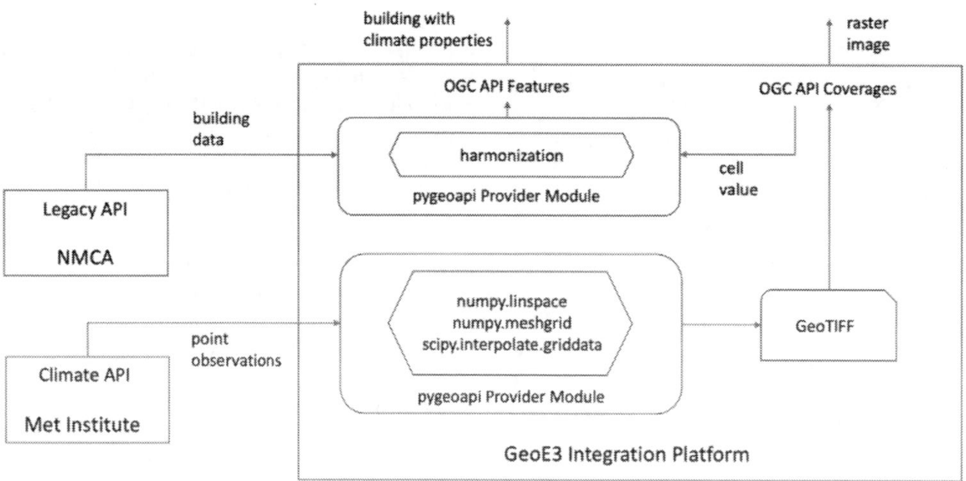

Figure 4. Processing pipeline for the OGC API Coverages-based provision of climate data on the GeoE3 integration platform.

Visualization System for the Positioning of Sunken Vessels Using Underwater Acoustic Devices

Dohyeong Kim Sang-Hyeok Nam*
Research Institute
ENGSOFT Inc.
Seoul, Republic of Korea
e-mail: dhkim@engsoft.kr e-mail: shnam@engsoft.kr

Jongryong Choi Donguoon Kim
Development Division Head Office
Ocean Planet Inc.
Busan, Republic of Korea
e-mail: xj2222@naver.com e-mail: daum5665@hanmail.net

Abstract— **In the event of a vessel sinking accident, it takes a lot of manpower, equipment, and time to search for lost vessels due to difficulties in securing underwater vision. If the searching work in the water is extended, excessive costs are incurred, and pollution of the marine environment may occur due to corrosion of ships and oil spills. In this paper, we introduce the visualization system for the underwater positioning using an underwater signal generator, a floating signal receiver and an underwater positioning algorithm to monitor the position of the sunken vessel. It can be minimized the human and material resources by maximizing the efficiency of underwater rescue and search operation for sunken vessels. In addition, it is expected that it can be used for various purposes such as submarine tracking, underwater monitoring and tracking the location of underwater rescue workers, underwater leisure sports, and so on through the minimizing of signal generation devices.**

Keywords-sunken vessel; acoustic transducer; floating signal receiver; underwater positioning algorithm; visualization system.

I. INTRODUCTION

This paper introduces the Sunken Vessel Position Identification System (SVPIS), which determines and visualizes the underwater position of a sunken vessels using an underwater signal generator and a floating signal receiver based on the short baseline (SBL) method as shown in Figure 1 [1][2]. As maritime traffic steadily increases worldwide, maritime accidents such as ship sinking continue to occur. When a ship sinks underwater, the ship may be lost by movement in the water by tidal currents or waves at the accident site. In addition, there are frequent cases of giving up the sunken ship due to excessive costs for the salvage of the ship. Wrecked ships whose salvage has been abandoned or failed are not identified visually at the water surface, which may cause secondary accidents with other ships, or cause marine environmental pollution due to corrosion of the hull or leakage of oil or chemicals [3]-[5].

A representative technology currently used to determine the location of the sunken ship is the Emergency Position Indication Radio Beacon (EPIRB) system. The EPIRB is deployed on the water surface of the sinking location and transmits the sinking location using a GPS signal. So, it is possible to know the initial location of the ship where the sinking accident occurred, but if the sunken ship moves underwater by tidal currents or waves, it cannot be tracked. In order to solve these problems, a method of tracking the location of a sunken ship by connecting a ship and a buoy with a cable may be used. However, this type has problems that cables can be cut by underwater floats or tidal differences and can be tangled during deployment in the event of an accident. Therefore, divers should be deployed to search the expected location of the ship, but it is difficult to identify and track the sunken ship due to the limited visibility, which takes excessive time and costs.

In SVPIS used in this work, the underwater signal generator emits an acoustic signal that is received by the floating signal receiver, which is floating on the surface of the water. The position of the sunken ship can be determined by utilizing the time differences of the acoustic signal arriving at each floating signal receiver and the GPS coordinates of the receivers. Once the position of the sunken vessel is determined, the SVPIS can visualize its location using a display unit in the visualization system. This allows the user

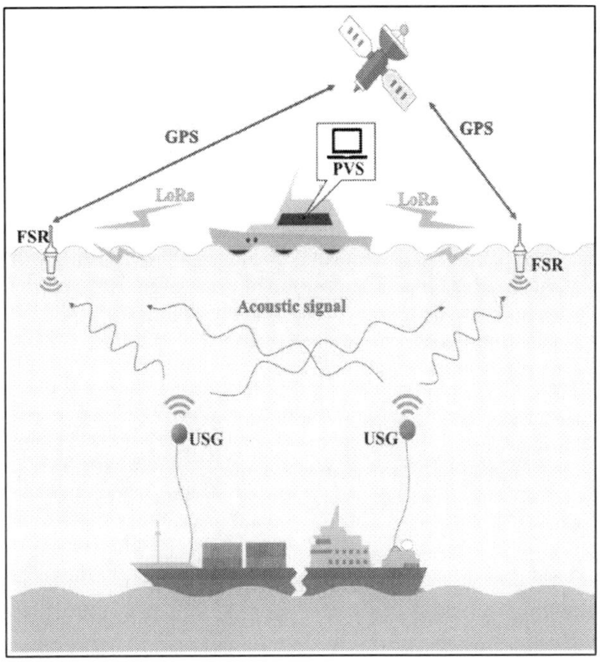

Figure 1. Overview of Sunken Vessel Position Identification System.

to see the precise location of the sunken vessel and plan any necessary recovery or salvage operations.

The rest of the paper is organized as follows: Section II introduces the components of the SVPIS and their respective roles, Section III describes the data flows necessary to visualize the position of the sunken vessel, on-site weather, and data measured by built-in sensors. Section IV details the underwater positioning algorithm using time difference of arrival to each Floating Signal Receivers, and shows the example of the visualization system for the sunken ship. And Section V summarize the lessons learned, conclusions and future work.

II. Components for the SVPIS

The components for the SVPIS include the Underwater Signal Generator (USG), the Floating Signal Receiver (FSR) and the Positioning & Visualization System (PVS).

A. *Underwater Signal Generator*

When the ship sinks, the USG attached to the hull is automatically deployed by water pressure and rises about 10 meters in the water while remaining connected to the cable, generating an acoustic signal. The USG contains an underwater acoustic transducer that can be attached to the bow or stern of the ship. The composition of the USG is illustrated in Figure 2.

B. *Floating Signal Receiver*

The main functions of the FSR are to receive acoustic signals from the USG while floating on the water surface and to transmit the received acoustic signal information with GPS information to the PVS. The FSR can be configured in various forms depending on the specific requirements, commonly including acoustic signal receiver modules, GPS modules, communication modules, and sensors (such as temperature, flow velocity, and turbidity sensors). Additionally, sensors may be added or excluded, or may also be manufactured in the form of a movable drone.

C. *Positioning & Visualization System*

The PVS comprises portable hardware that can calculate the location of the sunken ship using data obtained from the FSR and system software that visualizes the determined position based on GIS. The position of the sunken ship can be calculated based on the data received from the FSR via LTE or LoRa networks on rescue ships, the location (GPS coordinates) of each FSR, and sensor data. The accuracy of deriving the location of sunken ships will be improved through continuous research, including the application of noise filtering techniques, deep learning algorithms, and so on.

III. Design Concept of SVPIS

Figure 3 presents an overall schematic diagram of SVPIS, illustrating the data flow and connections between devices. In the event of a sinking accident involving a vessel, the USG operates through water pressure caused by water flowing into it, generating acoustic signals. The rescue boat deploys the FSR on the surface of the water, which receives sound wave signals emitted by the USG. The time of arrival (TOA) of the acoustic signal at the FSR, along with the GPS coordinates of each FSR and the data measured by the built-in sensors, is transmitted to the PVS via LoRa communication. Using positioning algorithms based on the TOAs and GPS coordinates of each FSR, the PVS calculates the position of the USG, which corresponds to the location (coordinates) of the sunken vessel. The PVS then visualizes the positions of the sunken vessel and FSRs, displaying measured data from the built-in sensors, and field weather data provided via the APIs, on the visualization system screen.

IV. Positioning and Visualization

The positioning algorithm of SVPIS calculates the difference in distance between each FSR and the USG by measuring the time difference between the times of arrival (TOA) of the signals received by different FSRs. This difference, combined with the known GPS coordinates of the FSRs, can be used to calculate the position of the USG, which corresponds to the location of the sunken vessel. In general, the time difference of arrival (TDOA)-based positioning algorithm uses either the "Least Square algorithm" [6] that obtains a solution by linearization or "Ho and Chan's algorithm" [7] that obtains a solution without linearization. In this work, the Least Square algorithm is used to derive the relative position of the USG by measuring the TDOA of each FSR. The absolute coordinates of the USG, which corresponds to the location of the sunken ship, are then calculated using the GPS coordinates of the FSRs as shown in Figure 4. However, in underwater environments, variations in sound wave velocity due to changes in water temperature or turbidity, as well as reflected sound waves by the water

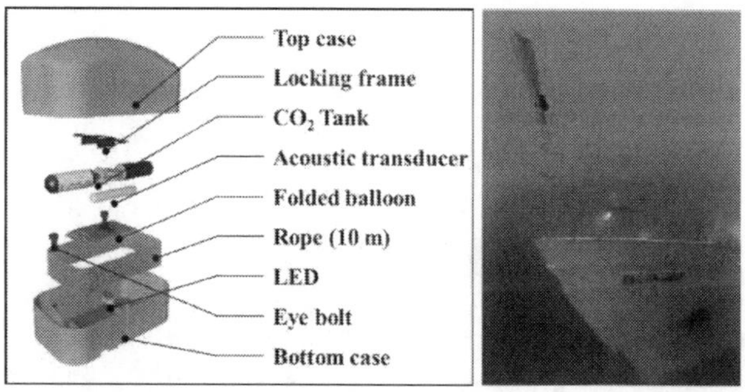

Figure 2. Composition of USG.

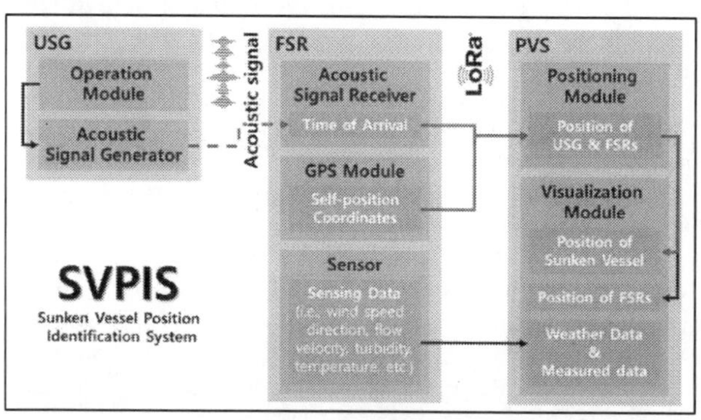

Figure 3. Schematic diagram of SVPIS.

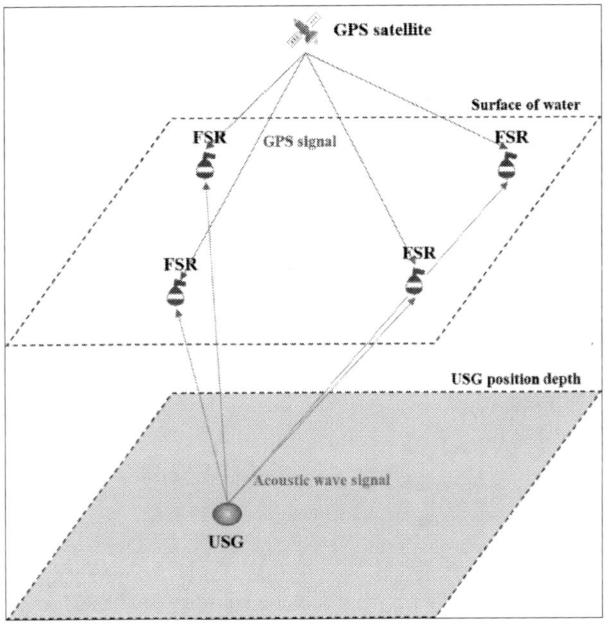

Figure 4. Underwater positioning.

surface or floating objects, can lead to inaccurate positioning results. Therefore, further studies are required to develop techniques that can increase the accuracy of underwater positioning by applying various technologies such as noise filtering technology, artificial intelligence modules, and others.

Figure 5 shows an example of a visualization system for tracking and monitoring the location of a sunken ship. The locations of the FSR and the sunken ship are indicated by icons on the GIS map, and the latitude/longitude coordinates and depth information of the sunken ship are informed by text. This information may be visualized on a three-dimensional map if necessary.

V. CONCLUSION

It has been introduced the Sunken Vessel Position Identification System, which uses underwater acoustic devices to rapidly and accurately calculate the position of a submerged vessel by measuring the time difference between the signals received by multiple FSRs in real-time. This allows for faster response times in rescue and recovery operations. The simplicity and efficiency of the SVPIS design, along with the use of relatively low-cost components, make it a cost-effective solution for underwater vessel positioning. Therefore, it has the potential to become a generic technology for responding to marine safety accidents through the positioning and tracking of sunken ships. Furthermore, improving the salvaging rate of sunken ships can reduce marine pollution caused by ship corrosion or oil spills.

In the near future, the SVPIS is expected to be used for searching for drownings, tracking the location of underwater rescue workers, and underwater leisure sports by minimizing the size of signal generation devices. To achieve this, continuous research is needed to improve the accuracy of underwater positioning technology and achieve a more efficient SVPIS.

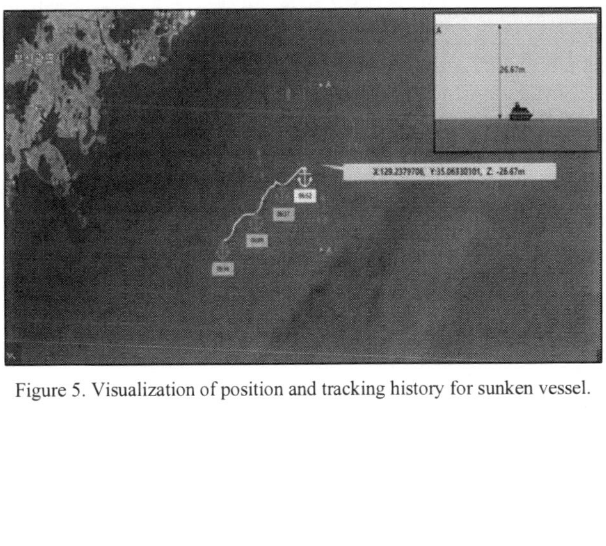

Figure 5. Visualization of position and tracking history for sunken vessel.

ACKNOWLEDGMENT

This research was financially supported by the Institute of Civil Military Technology Cooperation funded by the Defense Acquisition Program Administration and Ministry of Trade, Industry and Energy of Korean government under grant No. 2022DCME01.

REFERENCES

[1] P. H. Milne, "Underwater Acoustic Positioning System", E. & F.N. SPON, London, 1983.

[2] M. L. Neudoefer, "An Acoustic Position Reference System Integrating Long and Short-Baseline Operation", Proc. 11th Ann. Offshore Technology Conference, pp. 1721-1730, 1979.

[3] W. J. Chang, S. H. Lee, H. J. Yeom, and I. C. Lee, "A Study on the development of risk assessment for sunken vessels using remaining-fuel estimations model," Journal of the Korean Society of Marine Environment & Safety, vol. 22, no. 1, pp. 90–97, Feb. 2016.

[4] J. Michel et al., "An issue paper prepared for the 2005 international oil spill conference: Potentially polluting wrecks in marine waters," International Oil Spill Conference, pp. 1–40, May 2005.

[5] N. Goodwyn, "Potentially polluting wrecks-UK risk database & it's applications," WOW II Conference, MD, USA, p.6, June 2011.

[6] Wade H. Foy, "Position-location Solutions by Taylor-Series Estimation," IEEE Transcations on Aerospace and Electronic Systems, AES-12, No. 2, pp. 187–193, Mar. 1976.

[7] Ho K, and Chan Y, "Solution and performance analysis of geolocation by TDOA," IEEE Transcations on Aerospace and Electronic Systems, Vol. 29, No. 4, October 1993.

User-Generated Voice Navigation Editing System Using Block-Type Visual Programming Language

Daisuke Yamamoto
Nagoya Institute of Technology
Nagoya, Japan.
email:daisuke@nitech.ac.jp

Hiroki Sekiya
Nagoya Institute of Technology
Nagoya, Japan.
email:h.sekiya.976@nitech.jp

Shinsuke Kajioka
Nagoya Institute of Technology
Nagoya, Japan
email:kajioka@nitech.ac.jp

***Abstract*–Voice navigation systems on smartphones have become widespread in recent years. However, existing systems only provide one-way voice navigation based on uniform, machine-generated navigation instructions. It would be beneficial to have tourism navigation and two-way voice interaction, alongside normal navigation, for use in tourist hotspots. It is difficult to automatically generate interactive sentences for use in dialogue with map data alone. Therefore, this study aimed to create an environment in which users with knowledge of a given area can easily create voice navigation content. Google Blockly, a widely used block-type visual programming language, is extended and matched with map nodes (called "spots" in this paper) and links to produce a mechanism for achieving navigation at the spot level. Furthermore, a two-way voice navigation system is achieved by implementing a mechanism to convert from user-generated content to a format for voice interaction systems. The usefulness of the proposed system was evaluated by conducting experiments with a prototype system.**

Keywords-geographic information systems; user-generated content; voice interaction systems; voice navigation.

I. INTRODUCTION

Voice navigation systems on smartphones, such as Google Maps [1], have become widespread in recent years. These systems provide voice prompts for directions and distances at important locations such as junctions. However, existing voice navigation systems only provide one-way navigation based on uniform navigation sentences that are automatically generated from a map database. It would be beneficial for tourism if there were interactive voice navigation systems to supplement normal navigation with additional information in tourist hotspots. Additionally, if additional information is provided when on roads that are difficult to understand or require care when walking, users would be able to reach their destinations safely and without getting lost.

Data on tourism navigation and supplementary navigation are generally not included in map databases. As a result, it is difficult to automatically generate these types of voice navigation from map databases. Therefore, we focused on technology related to User-Generated Content (UGC) [2]. UGC allows users to edit content such as dictionaries and maps, e.g., Wikipedia and OpenStreetMap [3]. Incorporating user-generated mechanisms into voice navigation allows users to edit content based on tourism navigation and supplementary information. As demonstrated by OpenStreetMap, UGC is a technology intimately connected with the field of geographic information systems, and it has great potential.

Therefore, this study aimed to develop a mechanism based on UGC that allows users who are familiar with a given area to easily edit voice navigation scenarios ("voice navigation content") with two-way voice interaction functions. This enables voice navigation from departure to destination where it feels as though a pedestrian is listening to a tour guide.

The following requirements should be met to achieve the above objective.

Requirement 1. The mechanism, based on the concept of UGC, should allow anybody to easily edit two-way voice navigation content.

Requirement 2. In tourist hotspots, the shortest route is not necessarily the recommended route. The user should be able to change the navigation route.

Requirement 3. Pedestrians may get lost; therefore, it must be possible for the pedestrian to still be directed to their destination even if they do not follow the specified route.

In response to the above requirements, the proposed method has the following features.

Feature 1. We propose spot blocks, an extension of Google Blockly [4] matching blocks to map nodes (called "spots" in this paper). This allows editing of voice navigation content using a block-based visual programming language.

Feature 2. The navigation route can be edited easily by changing the combination and layout of spot blocks.

Feature 3. Setting the default state of the spot blocks to the shortest route enables the handling of brand-new routes.

Google Blockly is a technology that is widely used in Scratch [5], a visual programming environment for children.

There have not been many studies attempting to achieve voice navigation content based on UGC, it is a field of research with great potential. The results of our study contribute to the development of research fields related to geographic information systems, voice interaction systems, and user generated content, as well as the confluence of these fields.

In Section 2, we describe the related work. Section 3 describes the proposed method and Section 4 describes the proposed system. Section 5 describes the experimental results. Section 6 concludes.

II. RELATED WORK

Many studies on UGC have been conducted in the field of geographic information systems. For example, the following are studies on methods for collecting, analyzing, and applying geographic information from social media. Giradin et al. proposed a method for collecting spatio-temporal information about tourists by analyzing UGC [6]. Li et al. proposed a system called VisTravel, which collects, visualizes, and makes analyzable the opinions of tourism networks through UGC [7]. Khoshamooz et al. proposed a mechanism for extending the parameters for multi-criteria route planning utilizing UGC [8]. Hu et al. proposed a method for mapping the brand positioning and competitive situation of hotel brands by text mining of UGC [9].

There are several proposals for navigation based on the concept of UGC. Holone et al. proposed a mechanism based on UGC, which weights the routes navigated by the user themselves using a smartphone [10]. Yanagi et al. implemented voice navigation using tagged information that was posted on Twitter [11]. However, these studies used one-way voice navigation, unlike the proposed method which uses two-way voice navigation.

A system that achieves two-way voice interaction is called a voice dialogue system. Many studies have been conducted on voice interaction systems. MMDAgent (used in this study) is a system that combines the functions needed for achieving a voice interaction system, such as voice synthesis, voice recognition, 3D model drawing, and interaction control [12]. Dialogue scenarios can be formatted as Finite State Transitions (FST). Nishimura et al. proposed a framework for easily creating and sharing web-based scenarios for voice interaction systems based on the concept of UGC [13]. Wakabayashi et al. enabled the creation of content for voice interaction by manipulating the state

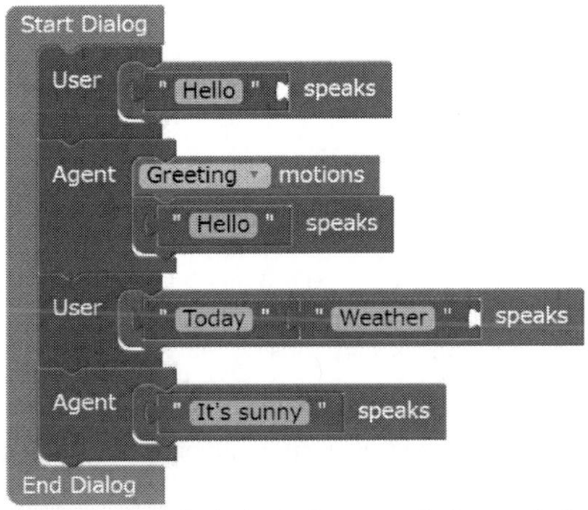

Figure 1. Example of voice interaction content blocks created using Google Blockly.

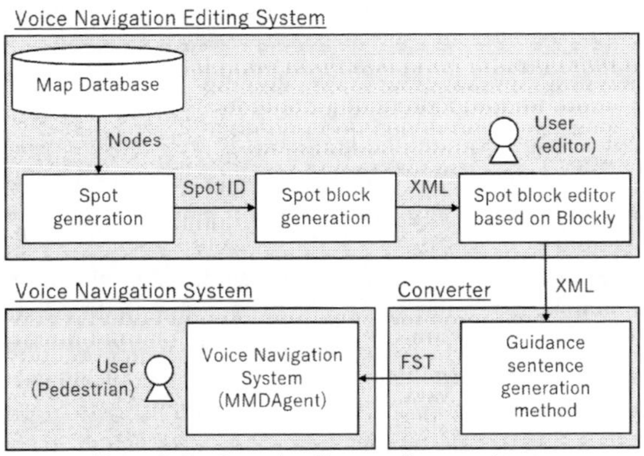

Figure 2. Configuration diagram of proposed system.

transition diagram using a tablet screen [14]. Additionally, Furuichi et al. used Google Blockly blocks to define voice interaction systems (voice recognition block, voice synthesis block, etc.) [15]. As shown in Figure 1, combining these blocks enables the creation of two-way voice interaction content. In this example in Figure 1, if a user says "Hello," the agent will respond with "Hello" while bowing. If the user asks "What is the like weather today?," the agent will respond with "It is sunny." These studies were conducted on generic voice interaction systems, and they do not deal with navigation. Hayashi proposed a method for editing voice navigation content [16]. The voice interaction content could be edited in the form of state transition diagrams in association with maps by improving upon the work of Wakabayashi et al. in [14].

III. PROPOSED METHOD

In this section, we describe the configuration of the proposed system and the proposed method.

A. *System Configuration*

Figure 2 shows the configuration of the proposed system. The proposed system was implemented by extending the system of Furuichi et al. [15]. Specific extended functions are as follows. First, in the spot generation function, all intersections in the target area are acquired as spots in the route database. Acquired spots can be confirmed via the map interface. The spot block generator converts each spot into a spot block that is associated with a point on the map interface. By default, the generated spot blocks are connected along the route order from the departure point to the destination. The navigation route can be changed by manipulating the spot blocks using the spot block editing function. Furthermore, voice recognition and voice synthesis blocks can be inserted to enable the creation of two-way voice interaction scenarios. Finally, the navigation sentence generation function converts the information into a format that can be handled by the voice interaction system.

B. Spot Generation

In this study, a spot was the smallest unit for navigation and corresponded to an intersection node. As shown in Table I, a spot has an ID, latitude and longitude coordinates, a set of neighboring spot IDs, and the ID of the next spot on the route. Furthermore, the construction of a spot network enables the calculation of the distance and direction between spots, and the acquisition of information needed for navigation.

The procedure that the spot generation function follows is shown below. First, the latitude and longitude of the destination are input, then all intersections within an x km radius around the destination are searched, and all obtained intersections are set as spots. Next, the shortest paths from all spots to the destination spot are determined using Dijkstra's algorithm. The next spot to be navigated to can be determined by finding the shortest route; therefore, = the minimum amount of voice navigation was generated without using the spot block editing function (which will be described later).

C. Spot Block Editing

In the method described by Furuichi et al., there are voice recognition and synthesis blocks, but there are no blocks for navigation or location information. Therefore, we propose the spot block as a new block for describing navigation. Spot blocks correspond in a one-to-one manner with map spots.

TABLE I. DEFINITION OF A SPOT.

name	type	description
ID	int	Spot number
Latitude	double	Latitude
Longitude	double	Longitude
NextSpot	int	ID of next spot
NeighborSpot	int[]	IDs of neighboring spots

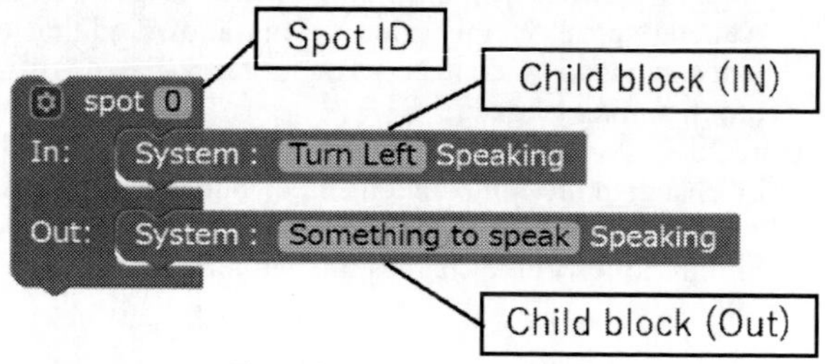

Figure 3. Example of a spot block.

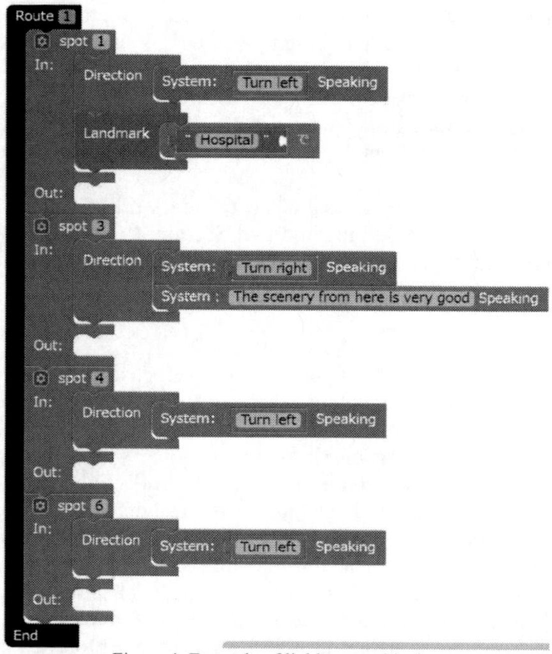

Figure 4. Example of linking spot blocks.

As shown in Figure 3, the spot block has a corresponding spot ID, a child block that fires when a user approaches the spot, and a child block that fires when the user leaves the spot. The spot blocks can be linked back-to-back with each other to determine the order of navigation.

The user can link spots to create their own route. Additionally, child blocks can be added to enable voice interactions at spots. Akin to those in the method from Furuichi et al. blocks that conduct complex interaction control (control blocks and function blocks) were used in the present study as well.

Spot blocks that are not linked are called orphan spot blocks. An orphan spot block works as a single spot block and fires when a user approaches the spot. The intersection to be navigated to next is determined based on the shortest route. The introduction of the orphan spot blocks enables navigation to the destination even if a route other than the one set in advance by the user was taken.

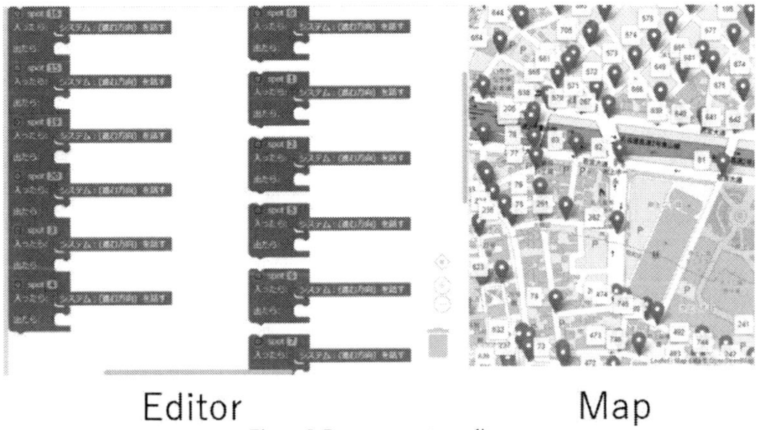

Figure 5. Prototype system edit screen.

Figure 4 shows an example of linked spot blocks. In this example, navigation is set in the order of spot ID 1, 3, 4, and 6, and when the spot ID approaches 3, supplementary interaction becomes possible with the voice prompt, "The view from here is very nice."

D. Navigation Sentence Generation

In the navigation sentence generation function, the block set generated by the spot block editing function is converted into a suitable format for the voice interaction system. The voice interaction system used is the Japanese voice interaction system construction toolkit MMDAgent.

The method for generating navigation sentences is the same as that of Furuichi et al., so the function for the voice navigation is explained here.

1) Generation of navigation sentences relating to directions and distance

Directing a pedestrian requires communicating the direction to turn (turn right, turn left, go straight, etc.) and the distance to the next spot. Therefore, if the ID of the spot where the pedestrian is currently located is S_1, the ID of the next spot to advance is S_2, and the ID of the immediately preceding spot for the pedestrian is S_3, then the direction A and distance D are obtained by the following procedure.

1. Let V_1 be the vector from S_3 to S_1.
2. Let V_2 be the vector from S_1 to S_2.
3. A is determined using the angle between V_1 and V_2.
4. Let the length of V_2 be the distance D.

However, if S_3 cannot be defined, such as at the starting point, then A also cannot be defined. In this case, navigation is provided using the direction of vector V_2 (e.g., north, south, east, or west).

2) Navigation route change function

The navigation route change function changes the route according to how the user interacts with the spot block. As previously mentioned, the user can determine the next spot to head toward by combining spot blocks. This is processed as follows according to the spot block operation event.

1. When spot block S_2 is connected to spot block S_1, the NextSpotID of the spot is the ID of the spot corresponding to S_2.
2. When the spot block S_2 is separated from spot block S_1, the NextSpotID of the spot corresponding to S_1 is set as the initial spot ID based on the shortest route.

Here, S_1 comes before S_2.

E. Prototype System

Figure 5 shows a prototype system based on the proposed method. In the prototype system, spots are displayed as blue pins on the map. The user edits the spot block using the spot block editing function while referring to the displayed spot. The edited blocks are automatically converted into the FST format for voice interaction systems, which enables the creation of voice navigation scenarios. Additionally, the created navigation scenario operates on a smartphone, and interactions related to voice navigation are possible according to the location and interaction content.

IV. Experimental Results

We conducted three experiments to evaluate the effectiveness of the proposed method. OpenStreetMap was the source of the map data.

A. Experiment on Navigation Sentence Generation

The proposed system should provide simple voice navigation based on Dijkstra's algorithm without needing to use the spot block editing function. Here, we verify if this function is operating effectively.

We use the prototype system to determine 10 destination points, acquire intersections, and place spots in the range of 500 m, 1 km, 1.5 km, and 2 km around the destination. We verify at this time whether it is possible to reach the destination by the route that is automatically obtained by Dijkstra's algorithm from an arbitrary spot. Table II shows the average arrival rate for each distance. The arrival rate exceeded 98% for all distances.

These results showed that a minimal amount of voice navigation was necessary even without user editing. This allows the user to concentrate on editing such as tourism navigation and supplementary navigation.

B. *Experiments on the Editing Interface*

Next, we tested the interface of the spot block editing function based on the block-type visual programming language.

We gave eight university students the task of creating an interactive route for voice navigation between two points using both the prototype system and a conventional method. We measured the time for creation and conducted a survey after the experiment. We used the system usability scale (SUS) [17] as a usability evaluation scale. The SUS score has a maximum of 100 points and an average of 68.

TABLE II. AVERAGE ARRIVAL RATE FOR EACH AREA.

	500 m	1 km	1.5 km	2 km
Average arrival rate (%)	98.75	99.01	99.35	99.81

TABLE III. EXPERIMENTAL RESULTS ON EDITING INTERFACE.

	Hayashi's method	Proposed method
SUS Score	45.5	61.5
Creation time	170 seconds	157 seconds
Waypoint	4.8	6.2

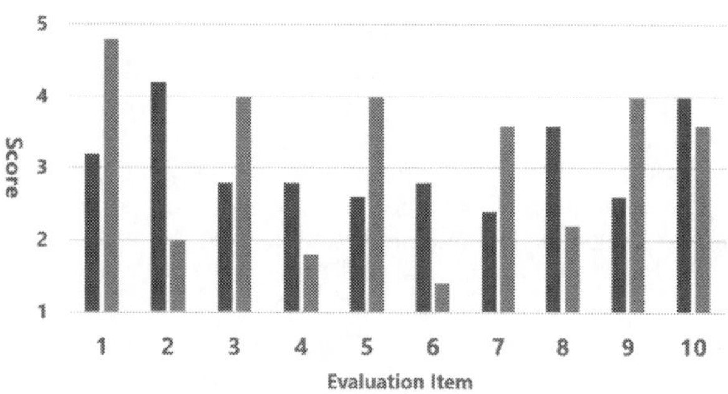

Figure 6. Details of SUS score relating to editing interface. Left: Hayashi et al. method, right: proposed method.

The systems to be compared are the following two methods.

1. Proposed method.
2. Method of Hayashi [16].

In the prototype system, there were 44 spots including the start and destination points.

Table III shows the experimental results. Figure 6 shows the average score for each item of the SUS score. In this figure, for odd-numbered items, the higher the number, the higher the rating. For even-numbered items, the lower the number, the higher the rating.

The proposed method scored the highest across all metrics of the SUS score. The conventional method scored 45.5, and the proposed method scored 61.5. This result suggests that the block-type description format of the proposed method is also effective in creating voice navigation scenarios. However, this score is lower than the score of the interaction scenario editing system by Furuichi et al., which attained 67.8. We think the reason for this is that the operation of the proposed system has become more complex due to the added functions for navigation. Therefore, further improvement of the interface is likely needed.

C. *Navigation Scenario*

Next, we verified the usefulness of the voice navigation scenarios generated by the prototype system.

Nine university students participated in this experiment. The systems to be compared were:

1. Automatically generated simple voice navigation (assuming Google Maps).
2. Method of Hayashi et al. [16].
3. Proposed method.

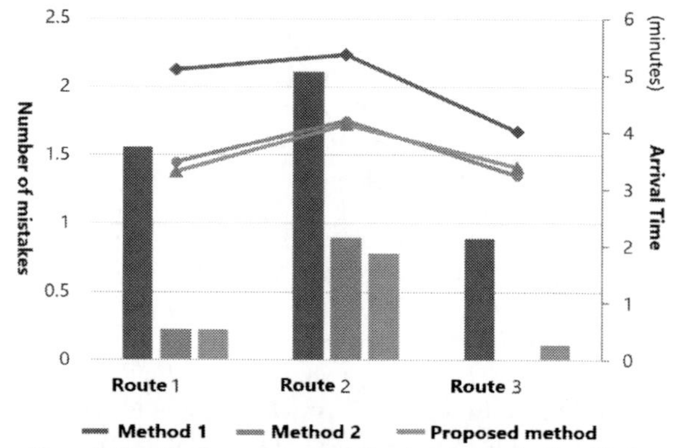

Figure 7. Results of experiments relating to voice navigation. The bar graph shows the number of mistakes, and the line graph shows the arrival time.

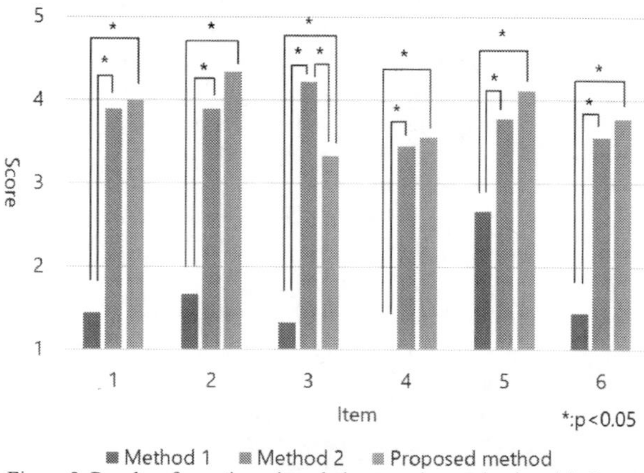

Figure 8. Results of questionnaire relating to voice navigation. * indicates a significant difference.

The automatically generated navigation sentences only gave instructions to turn at intersections. The navigation sentences generated by the method of Hayashi et al. supplemented the turning directions with landmarks. The navigation sentences of the proposed method included an interactive explanation of parts where users are likely to make a mistake to supplement the turning directions. The subjects used all three methods. The subjects were asked to record their arrival time and the number of times they made mistakes, and to respond to a five-level questionnaire that contained the following six items. Each level indicates how strongly the participant agreed with the phrase, a 5 indicated strong agreement.

1. Satisfaction with voice navigation.
2. I felt uneasy at corners.
3. I felt uneasy on the road.
4. I want to use this again.
5. The system speaks and responds in an appropriate manner.
6. Navigation is easy to understand.

Figure 7 shows the time required for each navigation and the number of times the wrong road was taken, and Figure 8 shows the results of the questionnaire.

The proposed method and that of Hayashi obtained better results than automatically generated sentences in the questionnaire and the experiments. For example, subjects often took the wrong turning when roads were narrow or at consecutive intersections. Such situations required nuanced navigation instructions such as "turn left in front of the cafe," instead of just "turn left at the cafe." Additionally, we think that the conventional method's arrival time was affected by the drivers checking carefully at turnings due to a lack of confidence about being on the right route. In the questionnaire, the conventional method and the proposed method scored significantly higher than the automatically generated method. In particular, items 2 and 3, uneasiness at junctions and on the road, were significantly reduced, which may have led to a higher overall satisfaction and an easier understanding of the navigation.

V. CONCLUSION

In this paper, we detailed the creation of a mechanism, based on the concept of UGC, that allows users with knowledge of a given area to easily edit voice navigation scenarios with two-way voice interaction functions. This allows for voice navigation from beginning to destination that feels like being guided by a tour guide. We extended Google Blockly by adding functions such as the spot block function.

We developed a prototype system based on the proposed method and discussed the effectiveness and problems of the proposed method. In the experiment relating to automatic generation of directions, the arrival rate from any spot in a range of 2 km centered on the target point was 98% or more. In the experiment relating to the editing system, the proposed method scored higher in the SUS and all questionnaire items; therefore, we conclude that the directions generated by the proposed method were easy to follow. When the directions were tested by test participants, it was found that they did not become anxious when using the voice navigation of the proposed method. Future work should include improving the UI of the spot block editing function and integrating multiple voice navigation content.

ACKNOWLEDGMENT

This work was supported by JSPS KAKENHI Grant Numbers JP19H04115 and JP21K19766.

REFERENCES

[1] Google Maps, https://www.google.com/maps/. [retrieved: 03, 2023]

[2] J. Krumm, N. Davies, and C. Narayanaswami, "User-Generated Content," in IEEE Pervasive Computing, vol. 7, no. 4, pp. 10–11, Oct.-Dec. 2008. doi: 10.1109/MPRV.2008.85. OpenStreetMaps.

[3] M. Haklay and P. Weber, "OpenStreetMap: User-Generated Street Maps," in IEEE Pervasive Computing, vol. 7, no. 4, pp. 12–18, 2008. doi: 10.1109/MPRV.2008.80.

[4] M. Seraj, E. S. Katterfeldt, K. Bub, S. Autexier, and R. Drechsler, "Scratch and Google Blockly: How Girls' Programming Skills and Attitudes are Influenced," Proceedings of the 19th Koli Calling International Conference on Computing Education Research, No. 23, pp. 1–10, 2019. https://doi.org/10.1145/3364510.3364515

[5] M. Resnick et al. "Scratch: programming for all," Communication of ACM, vol. 52, no. 11, pp. 60–67, 2009. https://doi.org/10.1145/1592761.1592779

[6] F. Girardin, F. Calabrese, F. D. Fiore, C. Ratti, and J. Blat, "Digital Footprinting: Uncovering Tourists with User-Generated Content," in IEEE Pervasive Computing, vol. 7, no. 4, pp. 36–43, Oct.-Dec. 2008. doi: 10.1109/MPRV.2008.71. Giradin.

[7] Q. Li, Y. Wu, S. Wang, M. Lin, X. Feng, and H. Wang, "VisTravel: visualizing tourism network opinion from the user generated content," Journal of Visualization, vol. 19, pp. 489–502, 2016. https://doi.org/10.1007/s12650-015-0330-x.

[8] G. Khoshamooz and M. Taleai, "Multi-Domain User-Generated Content Based Model to Enrich Road Network Data for Multi-Criteria Route Planning", Geogr. Anal., vol. 49, no. 3, pp. 239–267, 2017. https://doi.org/10.1111/gean.12124.

[9] F. Hu and R. H. Trivedi, "Mapping hotel brand positioning and competitive landscapes by text-mining user-generated content," Int. J. of Hosp. Manag., vol. 84, 102317, 2020. https://doi.org/10.1016/j.ijhm.2019.102317.

[10] H. Holone, G. Misund, and H. Holmstedt, "Users Are Doing It For Themselves: Pedestrian Navigation With User Generated Content," Proceeding of the 2007 International Conference on Next Generation Mobile Applications, Services and Technologies, pp. 91–99, 2007. doi: 10.1109/NGMAST.2007.4343406.

[11] A. Lee, K. Oura, and K. Tokuda, "Mmdagent—A fully open-source toolkit for voice interaction systems," 2013 IEEE International Conference on Acoustics, Speech and Signal Processing, Vancouver, BC, Canada, 2013, pp. 8382-8385, doi: 10.1109/ICASSP.2013.6639300.

[12] T. Yanagi, D. Yamamoto and N. Takahashi, "Development of mobile voice navigation system using user-based mobile maps annotations," 2015 IEEE/ACIS 14th International Conference on Computer and Information Science (ICIS), Las Vegas, NV, USA, 2015, pp. 373-378, doi: 10.1109/ICIS.2015.7166622.

[13] R. Nishimura, D. Yamamoto, T. Uchiya, and I. Takumi, "MMDAE: Dialog scenario editor for MMDAgent on the web browser," ICT Express, vol. 5, no. 1, pp. 47–51, 2019. https://doi.org/10.1016/j.icte.2018.03.002.

[14] K. Wakabayashi, D. Yamamoto, and N. Takahashi, "A Voice Dialog Editor Based on Finite State Transducer Using Composite State for Tablet Devices," Studies in Computational Intelligence, vol. 614. pp.125-139, 2015. https://doi.org/10.1007/978-3-319-23467-0_9

[15] M. Furuichi, D. Yamamoto, and N. Takahashi, "Voice Interaction Scenarios Editor with Block-based Visual Programming Facilities," Transaction on IPSJ, DCON, vol. 8, No.2, pp.1–15, 2020. (in Japanese)

[16] K. Hayashi, "Voice Dialog Scenario Editor Based on Spatial Extension of Finite State Transducer, " Master thesis, Nagoya Institute of Technology, 2019. (in Japanese)

[17] A. Bangor, P. T. Kortum, and J. T. Miller. "An empirical evaluation of the system usability scale," Int. J. Hum. –Comp. Int., vol. 24, no. 6, pp. 574–594, 2008.

Spatial Data Analysis as Decision Support Aimed at the Improvement of Agricultural Fertilization Strategies

Pundt, Hardy
Pleshkanovska, Roksolana
Harz University of Applied Sciences (HUAS)
Wernigerode, Germany
hpundt@hs-harz.de

Assmann, Denise
Böttcher, Falk
German Weather Service (DWD)
Leipzig, Germany
Denise.Assmann@dwd.de

Thiel, Enrico
Döhler, Johannes
Eißner, Florian
Kreuter, Thomas
Spott, Oliver
SKWPiesteritz (SKWP)
Cunnersdorf, Germany
Enrico.Thiel@skwp.de

Grunert, Michael
States Agency for Environment, Agriculture and Geology (LFULG)
Dresden, Germany
Michael.Grunert@smekul.sachsen.de

Abstract - Without fertilization, the sufficient provision of crops to feed the world population would not be possible. However, fertilization in agriculture also causes negative impacts. On the one hand, this concerns the environment e.g., by infusion of nitrate into the soil and groundwater. It is a challenge to guarantee both, a sufficient fertilization of soils, and the minimization of harmful emissions. The project "StaPrax-Regio", aims at an optimization of Nitrogen-(N)-fertilization strategies. It is well known that the interaction between weather, soil and the amount and time of fertilization are not independent from each other. Knowledge gaps still exist in a more detailed insight into the kind of interaction between the components, as well as the influence of specific parameters, such as precipitation and temperature, type, pH, cation exchange capacity of soils, and type, amount and time of fertilization. Therefore, meteorological and edaphic, as well as other factors must be taken into account when searching for a "good" fertilization practice at a specific site. Geographical Information Services (GIS) and Decision Support Systems (DSS) are adequate means to support data analysis and visualization, and finally the decision-making process on suitable fertilization strategies. The paper presents work-in-progress within the framework of a transdisciplinary research project of partners from industry, weather and geological services, and science.

Keywords. ***Fertilization, agriculture, soil data, weather data, GIS, spatial interpolation***

I. INTRODUCTION

Fertilizer nitrogen is mostly produced by applying the Haber-Bosch process [1]. Nitrogen (N) in the form of nitrate is a common pollutant in soils, surface and ground waters [2], [3]. Less than half of the more than 100 million tons of fertilizer N currently consumed yearly by agriculture is assimilated into the aboveground biomass of crops. While some fertilizer N will also be recovered by roots, much of the remainder is either leached or lost as environmentally harmful gas emissions [4]. Facing these facts, the goal of the StaPrax-Regio project is to achieve a better plant availability of N-fertilizers on the one hand, and to minimize harmful losses of N on the other. These goals were followed in other projects as well, however, StaPrax-Regio envisages a more comprehensive, data-driven approach thus integrating different information sources to achieve a holistic view on fertilization adapted as precise as possible to the local environmental conditions. The collaboration of different partners from practice, science and administration has been seen as a suitable way to achieve a multi-perspective, and therefore more sustainable, way to reach optimal fertilization strategies. Therefore, an industrial partner, producing N-fertilizers based on innovative methods (the consortiums leader), and administrations, such as weather and geological services, as well as a university, are collaborating within the StaPrax-Regio project.

Basic aspects and results of the research carried out in the first project year are presented, among others, in [5]. The main research question, how short-term meteorological events, as well as the specific soil conditions in different regions have influence on the capability of crop plants to absorb N, is still in the center of interest. New fertilizers include inhibitors that lead to longer availability of the fertilizer in the soil layers where roots of crops grow. In such a way they foster a better plant availability. Which consequences result from this in view of meteorological conditions, specific soil properties and fertilization practices? Facing questions like this, the partners work on different activities that are presented in an overview in Figure 1. The figure presents the steps that have to be gone within the project, from data collection and –processing to decision supporting maps. In the next section the input data sets, coming from different sources, and their analysis and visualisation are presented.

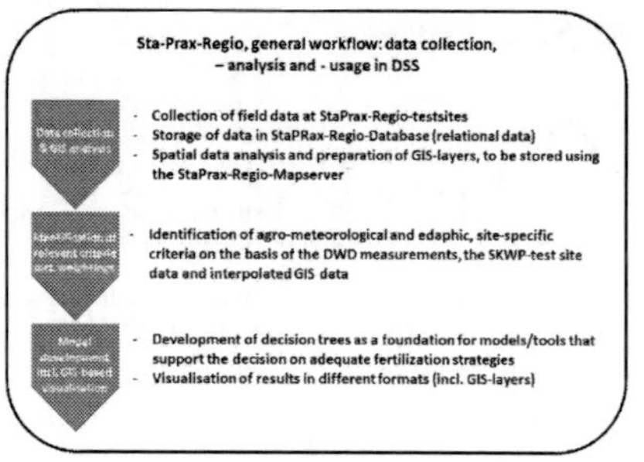

Figure 1. Data from agricultural test sites, meteorological stations and further sources have to be integrated to develop decision supporting models and tools [6]

The paper describes in the next section the current situation. Following, the data and the processing steps are explained that were carried out to achieve maps that are suitable to support decision making related to fertilization. The last section is aimed at mentioning further steps to be carried out during the next project year.

II. State of the Art

Fertilization in agriculture has ever been of large concern. Meanwhile, the subject has raised big attention: "(…) as our population grows to 10 billion people over the next few decades, we'll need even more nitrogen fertilizer. So, we find ourselves in a precarious situation: We need nitrogen for a healthy and growing population, but we're having a devastating effect on the environment by producing and using it." [7]. Much research is done concerning precision farming aiming at minimizing the amount of fertilizers used in agriculture. However, other attempts are on the side of the fertilizers themselves. An example is the stabilization with urease and nitrification inhibitors. This should lead to higher nitrogen efficiency and, as a follow-up, more environmental protection in agriculture. The urease inhibitor protects the urea from converting too quickly and also against ammonia losses. In addition to preventing N-losses, the subsequent nitrification inhibition leads to ammonium based plant nutrition, with adequate nitrate dispensation being assured at all times [15]. However, the dependence of fertilizing with such products from weather and soil conditions is obviously an important factor in successful fertilization strategies. Therefore, research attempts must proceed while taking into account more accurately all factors that play a role: meteorological and edaphic parameters, the various fertilizer alternatives, as well as further criteria [5], [8], [15]. Such a more holistic approach has been considered not sufficiently in the past. The purpose of the article is to inform about the methods and first results created during the still running project StaPrax-Regio.

III. Processing Data, Thus Preparing Them for Decision Support

A. Data acquisition and spatial analysis

Apart from official websites that provide spatial data, the partners Stickstoffwerke Piesteritz (SKWP), German Weather Service (DWD), and the States Agency for Environment, Agriculture and Geology of Saxony (LfULG) collect data at many sites throughout the country. The DWD carries out measurements at 67 different sites in all relevant landscape types of Germany, whereas SKWP collects data from 85 test sites. The States Agency provides foremost soil data, as well as data from own test sites.

In general, this data is point-based. Figure 2 shows an example of the results of the usage of different fertilizer combinations at one test site. The bars represent the results that have been measured at one and the same site while using varying quantities of different fertilizers (see abscissa) [9]. Such data have to be investigated in relationship to, among others, weather-dependent data [10], e.g., soil humidity (Figure 3).

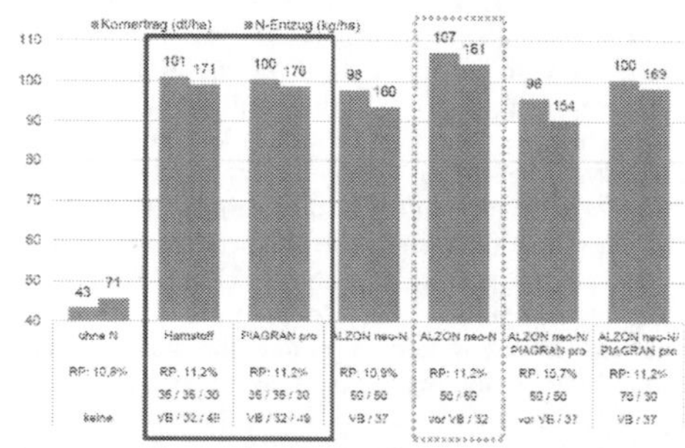

Figure 2. Exemplary, point-related results of fertilizer usage at a StaPrax-Regio-test site (Kornertrag = Grain yield; N-Entzug = Nitrogen removal) [9]

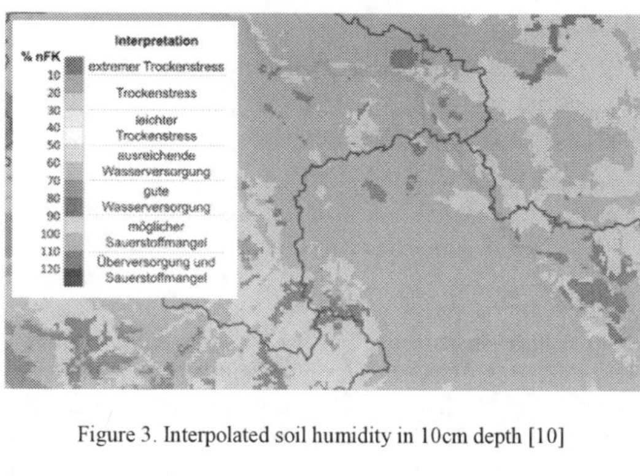

Figure 3. Interpolated soil humidity in 10cm depth [10]

For the provision of relevant data, the StaPrax-Regio MapServer has been implemented. The MapServer has been specifically configured for the StaPrax-Regio project and is based on a QGIS-Web-Client installation. It enables to publish GIS projects as OGC-compatible Web Map Services (WMS), Web Feature Services (WFS) or Web Coverage Services (WCS), and other formats. An adaptation of coordinate systems is easily possible and in such a way QGIS has been considered as an adequate and open-source solution for collecting, administering, analyzing and visualizing StaPrax-Regio related results. QGIS provides additionally functionalities and software extensions for spatial interpolation purposes. In such a way, the GIS tool includes important services to visualize, but foremost to modify the basic datasets and calculate area-wide layers which again can be overlaid with further relevant data. Concerning data quality, it can be stated that the positional accuracy is good enough for project purposes. The soil data relies on official data sources and the weather data are coming from the states weather service. All test sites that produce data on different fertilization strategies used are working on an approach that is acknowledged widely. Statistics based on the data applied in StaPrax-Regio are therefore robust and reliable.

B. Interpolation workflow

Most of the data acquired in StaPrax-Regio is point-related. The production of area-related information requires a suitable interpolation approach. A basic question to be answered here is the basic pattern in which the point data is provided [11]. Generally, there are three basic patterns in which such point data can occur, randomly (a), regularly (b) and clustered (c) (Figure 4).

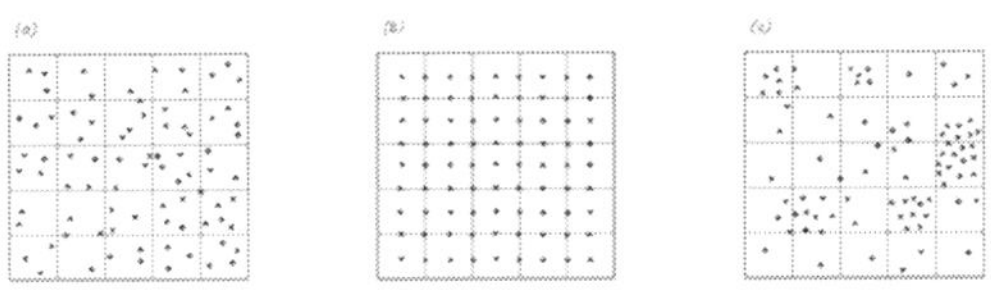

Figure 4: Possible patterns of point measurements [11]

For StaPrax-Regio, the data occur in a way that might be settled between (a) and (c). Distances between points vary as well as the number of points at different sites. These are spread over the whole country, however, different landscape types are represented through the test sites, the meteorological stations and the soil information. To achieve comparable results, an interpolation algorithm such as ordinary Kriging seems to be suitable to be used to generate area-related information on the various point-measured data. As mentioned before, the data used within the project are supposed to be reliable and therefore the interpolation results seem to be applicable for the overlay of interpolated and other spatial layers. Based on this assumption, the interpolation process is currently carried out on the usage of the software SmartMap, a QGIS extension. It provides, among others, the Kriging interpolation method [12]. Briefly summarized, the method relies on the semi - variogram which quantifies autocorrelation between the variance of data pairs and distance. At a certain distance autocorrelation becomes independent and this stage indicates that there is no longer any spatial relationship between the closeness of the data points. This makes Kriging to suitable linear, unbiased predictors under mild data conditions, thus proving to be a good method to interpolate spatial information [13]. Due to the irregular network of sampling points, the weather stations as well as the StaPrax-Regio test sampling sites (see section III.A and Figure 4), Kriging seems to be an adequate method, thus excluding non-relevant measurements, and having significant advantages concerning the given datasets and goal settings. Of course there are other interpolation approaches. It depends from the further project process if these should be tested and compared to the results achieved so far. Additionally, questions of isotropy and stationarity should be considered more sophisticatedly because edaphic and meteorological data might show differences here. Figure 5 presents the interpolation workflow as it is currently realized [14].

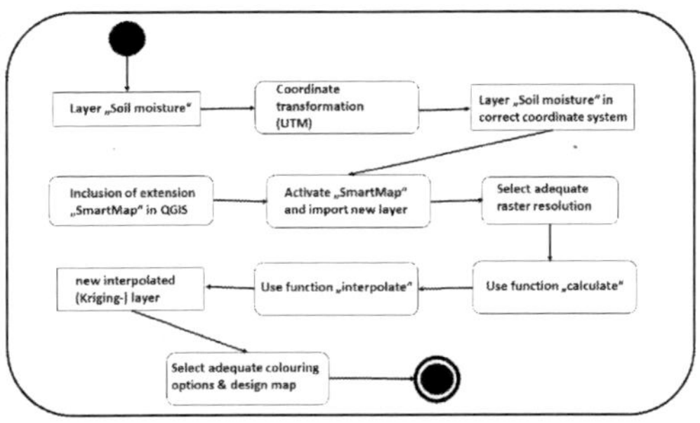

Figure 5. Exemplary GIS-based interpolation workflow [14]

C. Inclusion of processed data into decision-making procedures

Based on interpolated data, a "classic" GIS-overlay with further data layers can be performed which leads to new insights into local conditions (Figure 6):

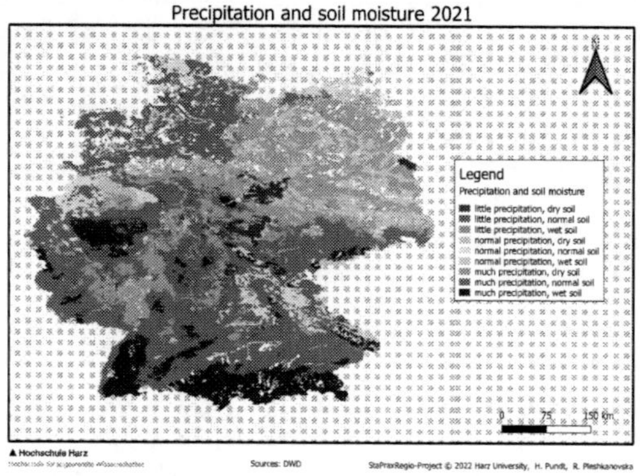

Figure 6. Overlay of previously interpolated data on soil humidity and precipitation [14]

The integration of the data from test sites, weather and edaphic information, including newly generated layers resulting from spatial analysis, will lead to spatial patterns which can help to improve decision-making on suitable fertilization strategies, especially envisaging a better uptake of fertilizers through crops under site-specific conditions [15]. This should have positive effects for the environment due to less emissions into the air, the (ground)water and the soil.

IV. CONCLUSIONS

The StaPrax-Regio-Mapserver provides basic spatial data layers, as well as analysis results, currently mainly based on classic methods such as spatial interpolation and overlay. However, the web-based system allows to explore the data in a multi-contextual view. This comes close to the insight that fertilization strategies should not consider only few factors, but the interaction between elements such as weather events, soil and landscape conditions, type of fertilizer and possibly machining technology. The overall project goal of such a holistic perspective on fertilization is to achieve more sustainable decisions instead of losing important information that can possibly, but not necessarily, play an important role [9].

After data collection at the test sites and the achievement of first results from analysis, the interpretation of such results is on the research agenda. Having defined procedures of how to prepare data for further analytical steps the project consortium will discuss adequate decision-making procedures. This requires differentiated thinking about the role each component plays and whether the various parameters can be handled uniquely or whether they have to be weighted in some manner. An efficient decision support will only occur when the spatial data layers become a helpful means to support farmers, agricultural consultants, and other decision makers. They should have helpful means to decide on a broad, comprehensive, and objective basis on a fertilization strategy that takes into account explicitly the edaphic and climatic particularities that vary from site to site. The definition of decision trees can help here [16]. This is also the basis for a DSS that is under development (web-BESyD [17]) at one StaPrax-Regio partner. It includes specific spatial datasets based on decision trees that are currently under critical review within the StaPrax-Regio consortium.

One important result is that defined different amounts of fertilizer combinations, including those with inhibitors, which were analyzed at all test sites, lead to different crop yields. This is due to the different edaphic and meteorological situations at the test sites. However, the overall goal is to answer the question whether spatial patterns can be identified, that allow to define recommendations, at best standardized fertilization strategies that are based on the data-driven research as it is carried out in StaPrax-Regio. Such strategies, taking into account explicitly all relevant components that play a role for crop growth and health, should be more environment-friendly, less expensive and therefore sustainable [5], [15], [18]. The first two project years gave important insights based on comprehensive spatial data form the test-sites that cover different landscapes throughout the country, thus enabling the integrated view on the situations. The next steps to be carried out will show, if such spatial patterns exist and if they deliver a basis to define best practices.

ACKNOWLEDGEMENTS

The StaPrax-Regio project is funded under the code 281DP02(A-D)20 by the Bundesanstalt für Landwirtschaft und Ernährung, BLE (State's Agency of Agriculture and Nutrition of Germany).

REFERENCES

[1] P. Patel, The Haber-Bosch Process: What Is It And Why Is The Process So Important? https://www.scienceabc.com/pure-sciences/the-haber-bosch-process-what-is-it-why-is-the-process-so-important.html, 2019. (accessed 2022-11-07).
[2] M. Sebilo, B. Mayer, B. Nicolardot, A. Mariotti, Long-term fate of nitrate fertilizer in agricultural soils. https://www.pnas.org/doi/10.1073/pnas.1305372110, 2013. (accessed 2022-11-03)
[3] A. Klinkenberg, The Agricultural Pollution of Europe's Groundwater. https://www.fairplanet.org/editors-pick/the agricultural-pollution-of-europes-groundwater/, 2022. (accessed 2023-02-26).
[4] U. Chrobak, The world's forgotten greenhouse gas. https://www.bbc.com/future/article/20210603-nitrous-oxide-the-worlds-forgotten-greenhouse-gas, 2021. (accessed 2023-02-18).

[5] H. Pundt, R. Pleshkanovska, E. Thiel T. Kreuter, F. Eißner, D. Assmann, F. Böttcher, M. Grunert, Optimization of Agricultural Fertilization Strategies based on Meteorological and Edaphic parameters: Analysis of Site-Specific Geoinformation to Support Decision Making, AGILE GIScience Ser., 3, 56, 2022.

[6] State's Agency for the Environment, Agriculture and Geology of Saxony (LfULG). Internal report, unpublished.

[7] B. Brady, Nitrogen Fertilizer: Solving This Environmental And Humanitarian Crisis. https://www.forbes.com/sites/ forbesbusinesscouncil/2022/05/12/nitrogen-fertilizer-solving-this-environmental-and-humanitarian-crisis/ (accessed 2023-03-13)

[8] S. Menegat, A. Ledo, R. Tirado, Greenhouse Gas Emissions from Global Production and Use of Nitrogen Synthetic Fertilisers in Agriculture. Scientific Reports volume 12, No 14490. https://www.nature.com/articles/s41598-022-18773-w, 2022. (accessed 2023-01-14).

[9] Stickstoffwerke Piesteritz (SKWP). Internal presentation on test-sites at StaPrax-Regio Meeting, Leipzig 2022-12-19, unpublished.

[10] German Weather Service (DWD), Soil Humidity-Viewer. https://www.dwd.de/DE/fachnutzer/landwirtschaft/appl/bf_view/_node.html (accessed 2023-01-14).

[11] A. Riha, Ein Vergleich räumlicher Interpolationsmethoden für Grundstückspreise im Bauland. University of Salzburg, Austria, Masterthesis, 2011, unpublished.

[12] Spatial Analysis, https://spatialanalysisonline.com/ HTML/ index.html?kriging_interpolation.htm (accessed 2022-12-27)

[13] P. Eguia, E. Granada, J. M. Alonso, E. Arce, A. Saav, Weather datasets generated using kriging techniques to calibrate building thermal simulations with TRNSYS. Journal of Building Engineering, Vol 7, pp 78-91, 2016.

[14] Harz University of Applied Sciences, Internal presentation of StaPrax-Regio Mapserver results at StaPrax-Regio-Meeting, Leipzig 2023-12-19, unpublished.

[15] F. Eißner, E. Thiel, F. Spott, F. Böttcher, D. Assmann, M. Grunert, H. Pundt, T. Kreuter, The collaborative project StaPrax-Regio – a central building block for location-based nitrogen fertilization (original title: Das Verbundprojekt StaPrax-Regio – zentraler Baustein für eine standortspezifisch optimierte Stickstoffdüngung). VDLUFA-Schriftenreihe 77, S. 231 – 234, 2021.

[16] J. Pearl, T. Wardly, Using a Decision Tree to help Make Decisions. https://extensionaus.com.au/ youngfarmersnetwork/ using-a-decision-tree-to-help-make-decisions/, 2019. (accessed 2022-11-29).

[17] LfULG, Further Development, Practical introduction and validation web-BESyD (original title: Weiterentwicklung, Praxiseinführung und Modellvalidierung web-BESyD). https://www.landwirtschaft.sachsen.de/weiterentwicklung-praxiseinfuehrung-und-modellvalidierung-web-besyd-43242.html, 2021 (accessed 23-03-2023)

[18] H. Pundt, Spatial Decision Systems. In: Leal Filho, W., Azul, A.M., Brandli, L., Özuyar, P.G., Wall, T. (eds) Climate Action. Encyclopedia of the UN Sustainable Development Goals. Springer, Cham, pp 837-848.

A Method for Removing Shadows from Photos Taken with a Drone and Stitching the Photos Together

YuJie Wu, Megumi Wakao, Naoki Morita
School of Information and Telecommunication Engineering
Tokai University
Tokyo, Japan
e-mail: 2cjnm024@mail.u-tokai.ac.jp, 9bjt2103@cc.u-tokai.ac.jp, morita@tokai.ac.jp

Kenta Morita
Faculty of Medical Engineering
Suzuka University of Medical Science
Mie, Japan
e-mail: morita@suzuka-u.ac.jp

***Abstract*—We are using a drone to obtain aerial views of grape farms to check the growth of grapevines. The shadows cast by the branches (cordons) in captured images differ depending on the location of the drone when the images were captured. The problem with stitching images of cordons with shadows together is that the cordons often appear disjointed at the edges of individual images. To solve this problem, we propose a method for removing elements other than cordons in the images captured from above, so that cordons only can be extracted from the images. We tested the effectiveness of this method using images in which the cordons were disjointed at the edges of the images after stitching by conventional methods.**

Keywords- Aerial image stitching; feature point; segment

I. INTRODUCTION

Viticulture in Japan is practiced by cultivating grapevines along trellises above the vines. Figure 1 shows a photograph of a vineyard, with branches(cordons) extending outward from the main trunk of the grapevines along a trellis positioned about 2 m above the ground. In order to harvest tastier grapes and manage nutrients effectively, it is necessary to monitor the number of shoots growing from the cordons [1]. However, since it is difficult to check cordon growth from the ground, we sought to use drones to check the state of the cordons from the air.

The characteristics of a grapevine are described here using aerial photographs obtained using a drone. Figures 2 and 3 show photographs of a grapevine taken from a height of approximately 10 m above the vines in a vineyard. The area of the image shown in Figure 3 was offset from the center of the area shown in Figure 2 by approximately 2 m. As indicated by the orange boxes in Figures 2 and 3, the lower half of Figure 2 and the upper half of Figure 3 show the same area. The feature indicated by the yellow arrow in Figure 2 is a cordon, the gray-white lines that are connected to a cordon are referred to as arms, and the brown line is also a branch. The black lines on the ground are the shadows cast by the vine. However, due to differences in the photographic angles, the appearance of cordon shadows is different. For example, the shadows of the same feature indicated by the blue arrows in Figures 2 and 3 are contained by the red circles in the two figures. Examination of the red circles shows that the distance between the cordon and the shadow is shorter in Figure 2 than in Figure 3. As a result of these differences, conventional image stitching methods could not be used to combine the images.

The aim of this study was to combine two images so that the cordons at the edges of adjacent images are correctly stitched together. Here we describe the stitching process that we developed to overcome the problems of disjointed cordons between images using the aerial photographs that we took.

Figure 1. Photograph of a grapevine in a vineyard in Japan.

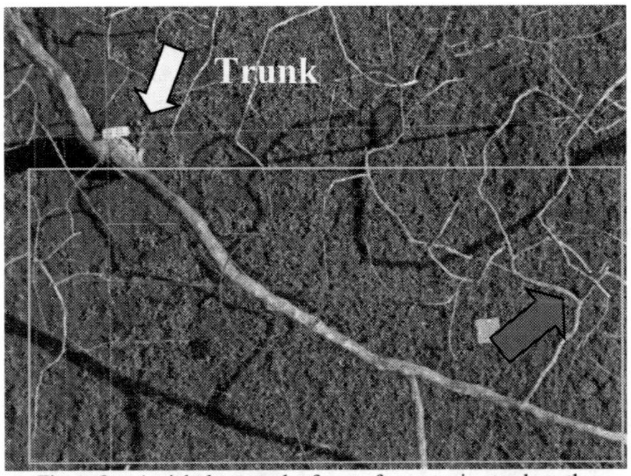

Figure 2. Aerial photograph of part of a grapevine to show the distribution of cordons and their shadows.

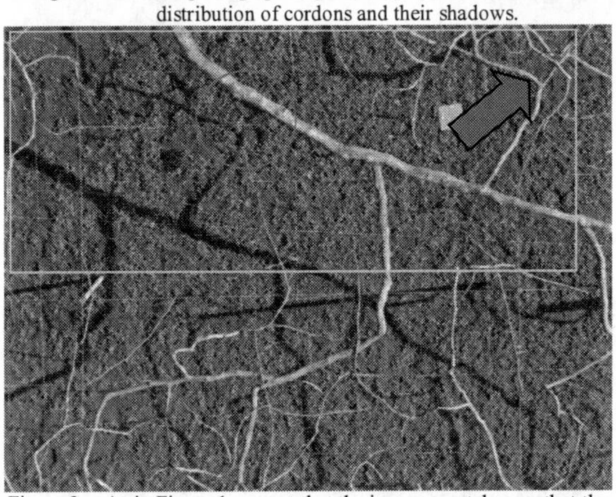

Figure 3. As in Figure 1, except that the image was taken so that the center of the images is offset by approximately 2 m.

In Section II we will explains the difficulties encountered with combining aerial photos and describes previous studies on the topic.

II. Difficulties In Combining Images

Images that were combined using photographs taken from slightly different locations showed that cordons in the image appeared cut off or disjointed. For example, Figure 4 shows the result of combining Figures 2 and 3 with the projective transformation using a Scale-Invariant Feature Transform (SIFT) descriptor. As shown by the red box in Figure 4, stitching of the two images resulted in the cordons and other features appearing disjointed.

In order to investigate the cause of these disjointed areas, we generated key points and descriptors using SIFT algorithm [2]. There are many methods for extracting feature values [3]-[5]. SIFT features are local, based on the appearance of the object at particular points of interest, and they are invariant to image scale and rotation. Moreover, the SIFT algorithm has a high recognition rate for feature points in regions with significant color shifts in the peripheral area. It is therefore considered to be particularly well suited for stitching together aerial images of orchards obtained in this study.

Figure 5 shows an image obtained by applying the SIFT algorithm to the central part of Figure 3. The circles in Figure 5 indicate SIFT key points and descriptors. The radius of the circle represents the feature scale. After checking the SIFT key points in Figure 4, the circles representing the feature points were generated for the entire image. Many of the key points identified were associated with features above the cordons. In addition, several circles were also associated with shadows, some of which were even larger than those associated with cordons and arms. This means that shadows are also being treated as features to be used for combining the image, i.e., just like cordons. Consequently, corrections need to be made to negate the effect of shadows when combining images of cordons.

Numerous studies have been conducted to correct distortions in terrain in aerial images [6]-[11], especially for buildings standing on the ground. However, the cordons are not attached to the ground. Orthoimages created from aerial photographs using GPS require reference points on the ground for correction. However, the "floating" nature of cordons makes it difficult to set reference points without including shadows, leading to significant image distortion. Conventional methods are thus not suitable for analyzing "floating" cordons.

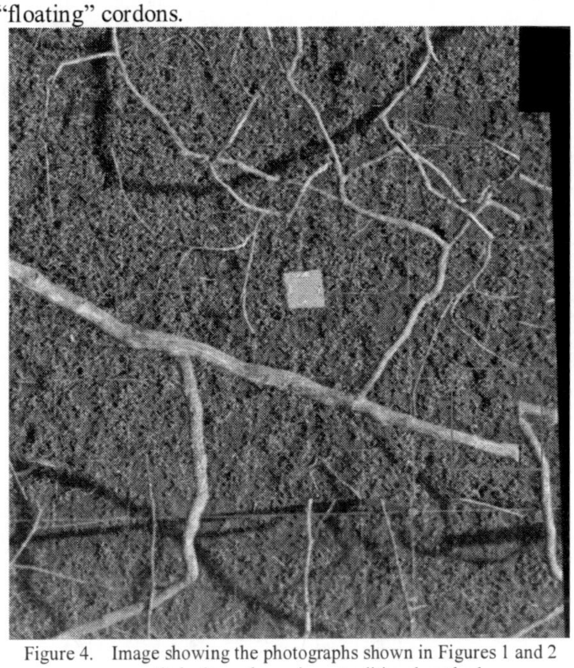

Figure 4. Image showing the photographs shown in Figures 1 and 2 stitched together using a traditional method.

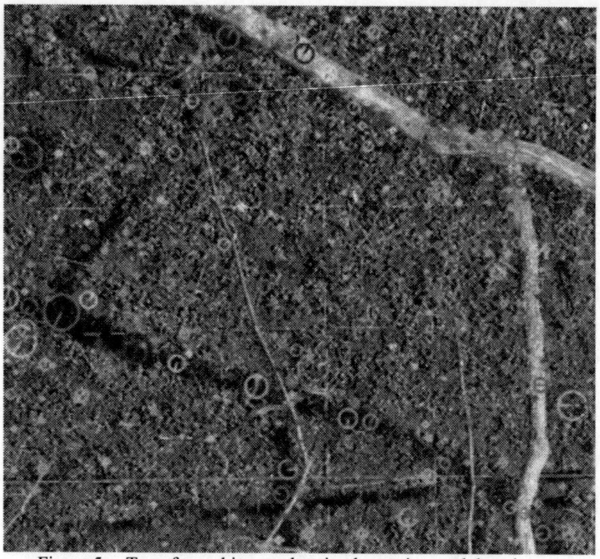

Figure 5. Transformed image showing key points and descriptors.

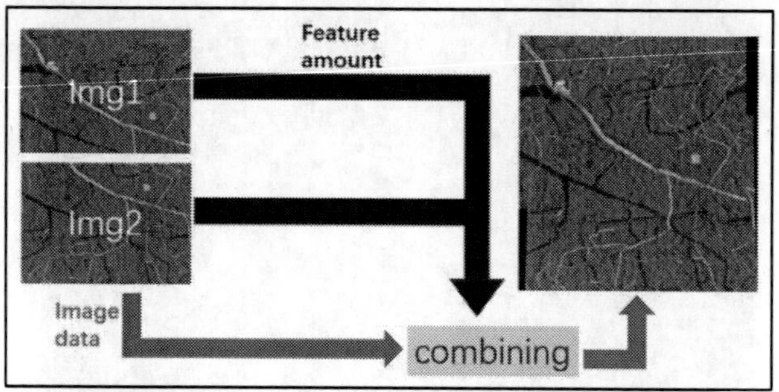

Figure 6. Conventional method for stitching images.

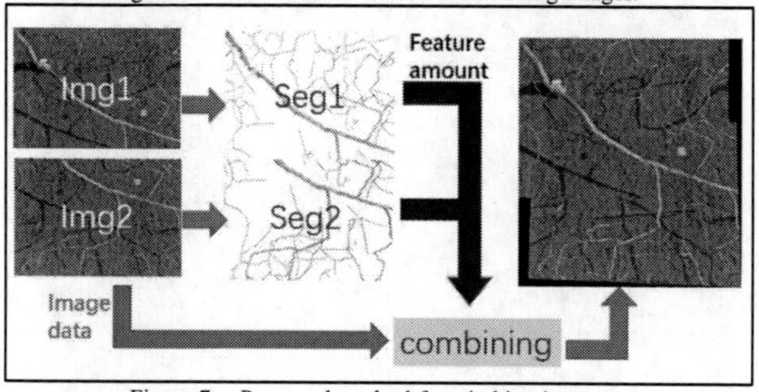

Figure 7. Proposed method for stitching images.

In Section III we will discusses the proposed solution.

III. PROPOSED METHOD

If the cordons and their shadows overlap each other when images are stitched, then the distortion increases. Therefore, we developed a method that ignores the effect of cordon shadows and stitches the photos so that only the cordons overlap each other.

Figure 6 shows a general overview of an image stitching procedure that uses feature points extracted from images to combine them. Figure 7 shows our proposed method. Our method first segments out the cordons from img1 and img2 to produce the seg1 and seg2 images that contain only cordons. Then, we apply the feature points for cordons extracted from seg1 and seg2 to img1 and img2 for image stitching. This method allows us to eliminate the effect of cordon shadows on image stitching and achieves seamless stitching of cordons at the edges of the images.

In Section IV we will describes the developed image-stitching system.

IV. SYSTEM DEVELOPMENT

To correct areas where the cordons are misaligned at the edges of aerial photographs, we developed an aerial photograph correction system. This alignment procedure can be divided into three steps.

Step1: Remove non-vine elements from aerial photos and create images that only show the silhouettes of cordons.

Step2: Extract feature points from the silhouettes.

Step3: Apply projective transformation to aerial photos based on the extracted feature points and stitch them together.

Using these three steps we created images that show only the silhouettes of the cordons. To remove the non-vine elements, we performed semantic segmentation using SegNet [12], a deep convolutional encoder-decoder architecture for image segmentation. We trained the network with 4000 images of cordons and segmented them with a network trained using 100,000 iterations. The segmentation results are shown in Figure 8. The reddish-brown regions shown in the figure represent the silhouettes of cordons.

Feature points and descriptors were extracted from the silhouette images using the SIFT algorithm. Figure 8 shows the SIFT feature points and descriptors obtained within the same region in Figure 5.

A projection transformation matrix was then computed based on the feature points extracted from the silhouette image, and the aerial photographs were merged. We employed the projection transformation function in the OpenCV library and stitched the images together based on the extent of matching among feature points.

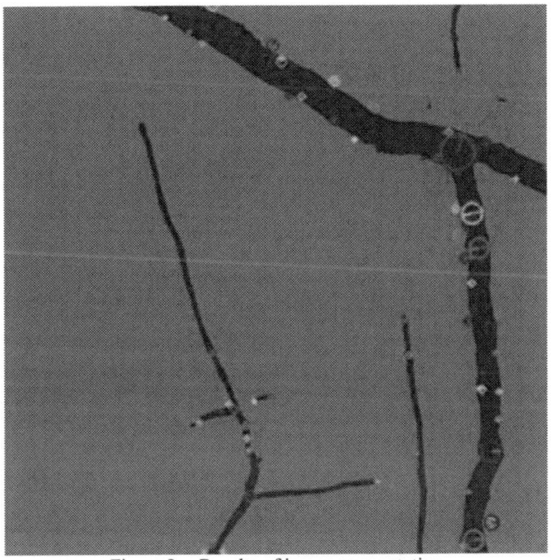

Figure 8. Results of image segmentation

In Section V we will describes the experiments conducted and presents the results and considerations, and the conclusions are given in Section VI.

V. EXPERIMENTS AND CONSIDERATIONS

In this section, we clarify the effectiveness of the proposed method. We compared the synthetic precision between the traditional method of stitching after extracting feature points from aerial images, and the method involving the extraction of feature points and stitching using images of cordon silhouettes.

Figure 9 shows one of the experimental results. It can be seen that the region enclosed in Figure 4 appears to be fully connected. As shown in Figure 4, the upper red box indicates a misalignment of four or more cordon widths, while the lower red box indicates a misalignment of three cordon widths. However, these defects are fully corrected in Figure 7.

Table I shows the measurement results. Using the conventional method, 53 cordons out of 60 were disjointed at the edges of images, whereas using the proposed method, only one cordon was disjointed. In the conventional method, one cordon was disjointed by a distance of twice the thickness of the original cordon, but using the proposed method, the cordon was disjointed by a distance of only one cordon thickness.

TABLE I. MEASUREMENT RESULTS

Disjointed distance	Traditional method	Proposed method
Nothing	7	59
One cordon	17	1
Two cordons	16	0
Three cordons	10	0
More	10	0

Compared with the conventional method, the proposed method improved the synthetic precision. Using the proposed method, only one cordon was not perfectly stitched. We considered that the reason for this discontinuity along the cordon was because, in terms of area, too little of the cordon was shared between the two original images being stitched.

From the experimental results, the method developed for recognizing cordons from aerial images and extracting feature points based on shared features was considered to be effective for connecting cordons seamlessly.

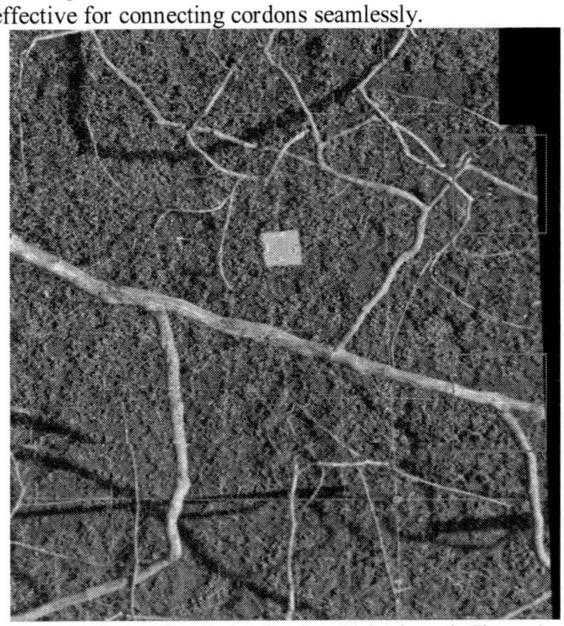

Figure 9. Image produced using the photographs shown in Figures 1 and 2 stitched together using the proposed method.

VI. CONCLUSION

The aim of this research was to accurately overlap cordons at the edges of aerial images being stitched. When combining photographs taken at two locations, cordons are often disjointed where the images are stitched. We therefore developed a method to remove elements other than cordons from aerial images and then extract feature points from images to show only the silhouettes of cordons. The findings showed that, compared with previous methods, the proposed method has a higher rate of synthetic precision for the stitching of cordons at the edges of aerial images.

In the future, we would like to stitch all of the aerial photographs so that the cordons overlap.

REFERENCES

[1] Nirasaki-shi, Yamanashi, CUPOD FARM [Online]. Available from: https://www.cupidfarm.co.jp/ [accessed: 2023-03-12]

[2] D. G. Lowe, "Object recognition from local scale-invariant features," Proceedings of the Seventh IEEE International Conference on Computer Vision, Kerkyra, Greece, 1999, pp. 1150-1157 vol.2, doi: 10.1109/ICCV.1999.790410.

[3] S. Leutenegger, M. Chli and R. Y. Siegwart, "BRISK: Binary Robust invariant scalable keypoints," 2011 International Conference on Computer Vision, Barcelona, Spain, 2011, pp. 2548-2555, doi: 10.1109/ICCV.2011.6126542.

[4] E. Rublee, V. Rabaud, K. Konolige and G. Bradski, "ORB: An efficient alternative to SIFT or SURF," 2011 International Conference on Computer Vision, Barcelona, Spain, 2011, pp. 2564-2571, doi: 10.1109/ICCV.2011.6126544.

[5] Alcantarilla, P.F., Bartoli, A., Davison, A.J. (2012). KAZE Features. In: Fitzgibbon, A., Lazebnik, S., Perona, P., Sato, Y., Schmid, C. (eds) Computer Vision – ECCV 2012. ECCV 2012. Lecture Notes in Computer Science, vol 7577. Springer, Berlin, Heidelberg. https://doi.org/10.1007/978-3-642-33783-3_16

[6] Guoqing Zhou, "Urban large-scale orthoimage standard for national orthophoto program," Proceedings. 2005 IEEE International Geoscience and Remote Sensing Symposium, 2005. IGARSS '05., Seoul, Korea (South), 2005, pp. 1214-1217, doi: 10.1109/IGARSS.2005.1525336.

[7] CAO Jian, LI Kan, GAO Chun-xiao, LIU Qiong-xin. Application of Local Features in Aerial Image Mosaic[J]. Journal of University of Electronic Science and Technology of China, 2013, 42(1): 125-129. doi: 10.3969/j.issn.1001-0548.2013.01.026

[8] Yamazaki Y., Itaya T. (2010) An attempt to create a wide-area ortho-mosaic image from analog aerial photographs. Journal of the Japan Society of Forest Survey, 32(4), 197-200.

[9] Kim, M., Kim, J., & Lee, D. (2018). A study on orthoimage generation method using unmanned aerial vehicle photogrammetry. Journal of the Korean Society of Surveying, Geodesy, Photogrammetry and Cartography, 36(6), 535-544.

[10] Guo, W., Wu, J., & Xie, X. (2019). Study on Ortho-image Mosaicing of Super-long-span Cable-stayed Bridge Based on Multi-sensors. Journal of Sensors, 2019.

[11] Ayhan, E., & Uysal, M. (2017). Estimation of stand height and aboveground biomass of Mediterranean pine forests using QuickBird-2 orthoimages. International Journal of Remote Sensing, 38(12), 3655-3677.

[12] Badrinarayanan, V., Kendall, A., Cipolla, R., "SegNet:A Deep Convolutional Encoder-Decoder Architecture for Image Segmentation." IEEE Trans. Pattern Anal. Mach. Intell. Vol 39, No.12, pp.2481–2495, Dec.2017

Indoor/Outdoor Route Estimation Method Based on Global Map Matching Using BLE Beacons and GPS

Shinsuke Kajioka*, Takanori Saito† and Daisuke Yamamoto‡
Department of Computer Science
Nagoya Institute of Technology
Gokisocho, Showa, Nagoya, Aichi 466-8555, Japan
*kajioka@nitech.ac.jp, †t.saito.536@nitech.jp, ‡daisuke@nitech.ac.jp

***Abstract*—In recent years, devices, such as smartphones and Bluetooth Low Energy (BLE) beacons, equipped with BLE functionality have become increasingly widespread, resulting in a proliferation of services that use BLE beacons to estimate location. However, most existing research on location estimation using BLE beacons is limited to indoor locations, and few methods are available to accommodate both indoor and outdoor locations. This study proposes a method that combines data from BLE beacons and Global Positioning System (GPS) to estimate continuous path of human movement across both indoor and outdoor areas. This enables not only highly accurate estimation of indoor and outdoor routes but also flexible route estimation by expanding "indoor" and "outdoor" routes to include "semi-outdoor" routes: for example, supplementary BLE beacons are placed near high-rise buildings or in semi-outdoor areas where GPS signals are weak to improve the accuracy of estimation. For achieving such estimation, the proposed method joints maps on adjacent floors or indoor/outdoor. To validate the effectiveness of the proposed method, a prototype system for achieving this functionality is developed and experiment is conducted. The results of the experiment indicates that the proposed method produces better results of route estimation.**

***Index Terms*—map matching, BLE beacon, GPS, geographical information systems**

I. Introduction

Devices, such as smartphones and beacons, equipped with Bluetooth Low Energy (BLE) functionality have become increasingly popular in recent years. As a result, BLE-based services, particularly location-estimation services using BLE beacons (hereafter referred to as "beacons") are steadily gaining popularity. As shown in Figure 1, the Nagoya Institute of Technology, to which the authors belong, has installed approximately 1,600 BLE beacons in all classrooms and hallways on campus [1], providing a smartphone service that can manage class attendance by receiving BLE beacon signals. BLE beacons are also expected to operate reliably even during power outages and disasters because of their low installation costs, stand-alone operation, low power consumption, and ability to operate for more than five years on dry cell batteries. Owing to these factors, studies estimating indoor location using BLE beacons have proliferated.

As a method of beacon-based location estimation, we have proposed a global map matching method [2] that uses beacons to estimate users' routes and stays. This is a route-and-stay estimation method that uses two networks, a route network for schematically representing a person's movement path and a beacon network for schematically representing the location relationship of beacons. By integrating and filtering beacon signals received using a smartphone, the system estimates users' routes and stays. On the contrary, most existing BLE beacon-based location estimation methods, including our method, are limited to indoor locations. BLE beacons have also received attention in estimating pedestrian traffic and can be used to analyze people flows for marketing, urban planning, evacuation drills, etc. For example, if there is an evacuation drill in an urban area, people will move to evacuation centers (parks, etc.) through various locations, such as building interiors, underground malls, and outdoors. If these indoor/outdoor routes could be estimated with high accuracy, this would contribute to better evacuation and urban planning.

This work was supported by JSPS KAKENHI Grant Numbers JP19H04115 and JP21K19766.

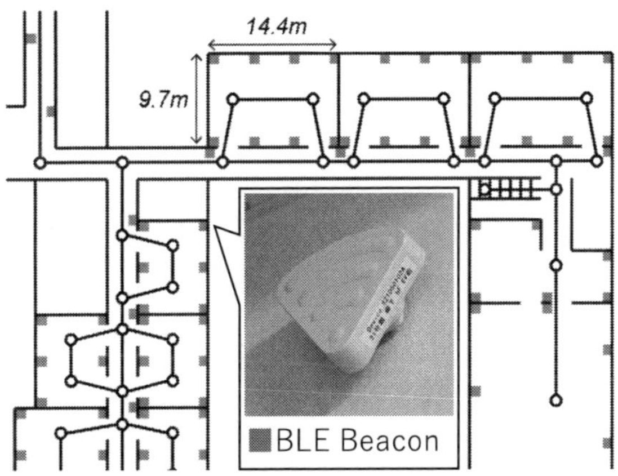

Fig. 1. A floor plan illustrated with locations of installed BLE beacons and candidate paths. A total of 1,600 BLE beacons are installed in all classrooms on campus and an attendance management system uses installed locations of BLE beacons.

At present, most research on location estimation using BLE beacons is for indoor use; however, there is little research on integration BLE beacons with Global Positioning System (GPS), which is used outdoors. Studies using BLE beacons typically estimate position considering the radio reception strength as input data, while studies using GPS typically estimate routes on the road network considering latitude and longitude as input data (routes are more abstract than a series of positions, and depending on the application, more useful). The combination of GPS and BLE beacons is inherently difficult because of the different data formats, methods, and estimation targets.

This study proposed a method that combines BLE beacon and GPS data to estimate an overall optimal path of human movement across both indoor and outdoor areas. This will not only enable highly accurate estimation of indoor and outdoor routes but also enable flexible route estimation by adding "semi-outdoor" to "indoor" and "outdoor" routes, for example, by placing supplementary BLE beacons near high-rise buildings or in semi-outdoor areas where GPS signals are weak, to improve estimation accuracy.

While it is standard practice to estimate a route after estimating location, we propose a method that estimates routes directly with high accuracy without estimating location. In other words, by specializing on route estimation, the system provides high accuracy and solves various problems that arise in practical use. This may contribute to the development of geographic information systems, including positioning information.

The rest of this paper is organised as follows: Section II shows related work on route estimation based on GPS data, Section III describes the definitions of the proposed system, Section IV discusses the proposed route estimation method, Section V shows evaluative experiment of the proposed method, and Section VI concludes this study.

II. Related Work

There has been significant research on route estimation based on GPS data. There are two main methods: one based on incremental map matching and the other based on global map matching.

Brakatsoulas et al. [3] provides an example of an incremental map-matching-based method. Two similarity indices, distance and orientation, were used to evaluate candidate links. Lou et al. [4] provides an example of a global map matching method. This study proposed ST-matching, which can perform high-accuracy map matching under low sampling rates with sampling intervals of 2 min or more. Score matching considers spatial and temporal scores and applies the Viterbi algorithm based on the scores for each candidate point and scores between candidate points. Newson and Krumm [5] proposed a globally optimal map matching based on a hidden Markov model.

In [6], a smartphone application was created to conduct a stamp rally at an event site, using booths scattered throughout the site as checkpoints. To realize user behavior analysis, they estimated indoor and outdoor routes using GPS and beacons. However, it lacked the high-level integration of the method proposed herein.

In [2], the authors conducted a study on beacon-based global map matching to estimate users' travel routes and stays. The authors proposed BST-matching, an improved version of ST-matching [4] for beacons, which determines candidate user locations at a given time based on beacon signal strength and uses the Viterbi algorithm for route estimation.

In [7], beacons were used for route estimation indoors, and GPS was used outdoors where beacons were not installed to extend the scope of possible route estimation.

Ito and Kawaguchi [8] described the process of estimating a user's indoor and outdoor travel paths using the optimal available positioning technology for the surrounding environment in which the user was moving. By defining in advance which positioning technologies are available in the road network and switching the information used for estimation, it becomes possible to use complex information for route estimation.

Several methods based on machine learning have also been proposed. Xiao et al. [9] proposed a method for location estimation using BLE beacons in 3D space using Auto Encoder. Urano et al. [10] proposed a method for indoor BLE beacon location estimation based on Long Short-Term Memory (LSTM). Unlike our work, these do not mention the fusion of BLE beacons and GPS.

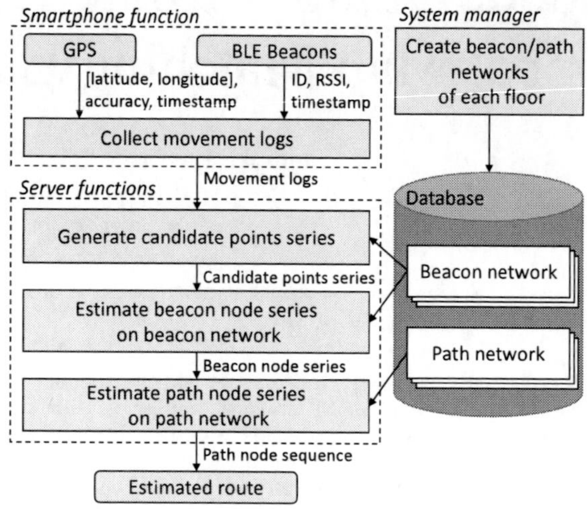

Fig. 2. Route estimation procedure of the proposed system. A smartphone function collects both GPS data and BLE beacons as movement logs. Server functions process the movement logs and derive a path node sequence as an estimated route. System manager should create beacon/path networks beforehand.

III. Proposed System

This section presents the definitions of the movement log and the path and beacon network.

A. *Configuration of the Proposed System*

Figure 2 shows the procedure structure of the proposed system. System manager should prepare beacon and path networks of each floor or exterior beforehand. First, the log collection function collects movement logs from GPS and BLE beacon signals on a smartphone. The collected movement logs are stored and carried them to the route estimation server.

TABLE I
BEACON LOG FORMAT.

ItemName	Type	Description
Beacon_ID	integer	Beacon ID
RSSI	double	RSSI
Timestamp	ISODate	Timestamp

TABLE II
GPS LOG FORMAT.

ItemMame	Type	Description
latlon	Position	Location coordinates expressed as latitude and longitude
Accuracy	double	Accuracy
Timestamp	ISODate	Timestamp

The route estimation server converts the movement logs into a time-series of sets of candidate points, which is then converted into a series of beacon nodes in the beacon network. Subsequently, the series of beacon nodes on the beacon network is transformed into a series of path nodes on the path network, and the estimated route of user movement is output.

B. Movement Logs

In this study, to estimate routes, BLE beacon data and GPS data are stored as movement logs using smartphones carried by users.

The movement log has two fragments of information in chronological order: the beacon log and GPS log. The beacon log format is presented in Table I, and the GPS log format is presented in Table II. The beacon log consists of the beacon's ID, Received Signal Strength Indicator (RSSI), and a timestamp. The GPS log consists of latitude and longitude coordinates, accuracy, and a timestamp.

C. Path and Beacon Network

Similar to our previous method [2], this study uses two networks, path and beacon, as shown in Figure 3. A beacon network represents beacon installation as a graph, with nodes located at beacon positions. The path network represents the actual candidate routes that users will take, regardless of the beacon position.

The conventional method deals only with indoor path and beacon networks, whereas the proposed method necessitates the construction of both indoor and outdoor networks. In this study, indoor and outdoor networks were created separately and connected using stairway nodes and ingress/egress nodes, as described below. These nodes function as connecting nodes in the conventional method [2].

The network is edited using our previously developed indoor/outdoor network editing system [11].

Stairway nodes are used to connect building floors. An example of a stairway node is shown in Figure 4. The corresponding stairway nodes on each adjacent floor have the same coordinate. They are installed at the landing of a staircase in an indoor map and have the effect of connecting maps on adjacent floors.

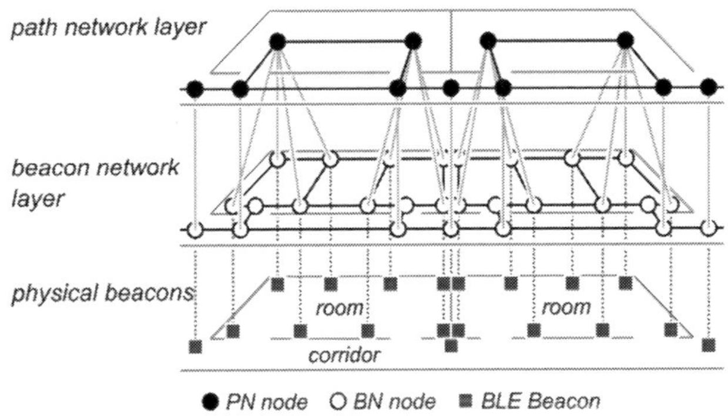

Fig. 3. Path and beacon network. "PN node" and "BN node" indicate a node on path network layer and beacon network layer, respectively.

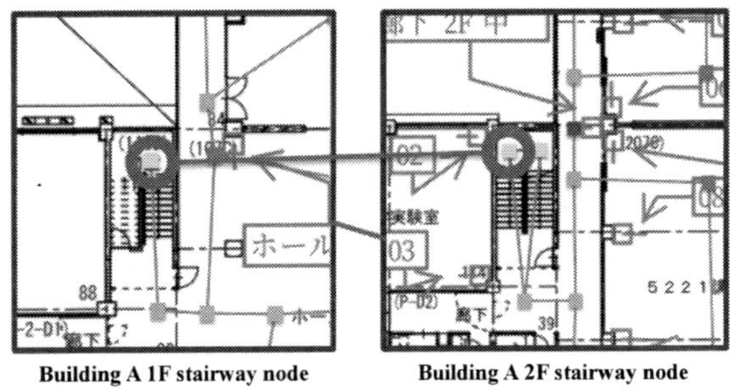

Fig. 4. Example of a stairway node. The red circle indicates the stair node connecting the first and second floors.

Ingress/egress nodes are nodes that connect indoor networks to outdoor networks. An example of corresponding ingress/egress nodes are shown in Figure 5. The corresponding ingress/egress nodes on the same location of each map have the same coordinate. Ingress/egress nodes are positioned between building interiors and exteriors to connect corresponding maps on adjacent indoor/outdoor.

IV. ROUTE ESTIMATION METHOD

In this section, we propose a method to estimate indoor and outdoor routes taken by users using GPS and BLE beacon movement logs as well as the path and beacon network.

A. Combination and Filtering of Movement Logs

In general, BLE beacon signals and GPS signals are prone to interference from buildings and obstacles. Therefore, their accuracy is likely to vary based on location and time of day.

Therefore, in this section we present a method for combining and filtering movement logs as a preprocessing step before performing the estimation.

When combining movement logs, logs are arranged in chronological order from beginning to end, and from the resulting divided i logs, only the information with the highest accuracy is retained. The combination process is shown below: Note that $i = 3$ in this study.

1) The movement log was split into i parts in chronological order.

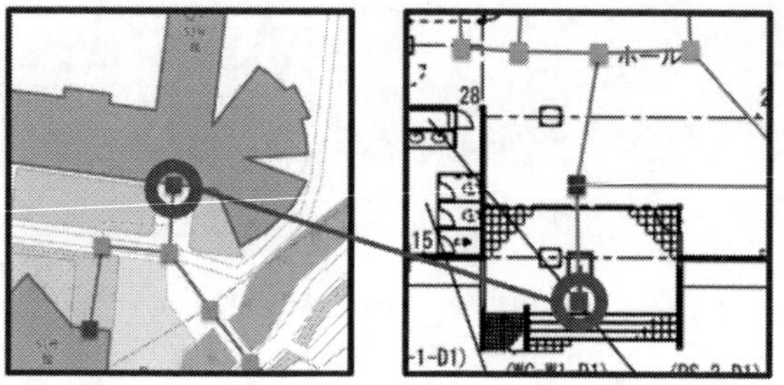

Fig. 5. Example of an ingress/egress node. The red circle indicates the ingress/egress node connecting the outdoor and indoor.

2) If there were multiple beacon logs with the same major value among the divided i beacon logs, the one with the highest RSSI value was selected.
3) From the separated i GPS logs, the log with the highest accuracy was selected.
4) The i movement logs were combined, and the time after combination was considered to be the newest time.

Subsequently, we explain filtering: Filtering removes movement logs below a specified threshold from the combined movement logs. The filtering procedure is shown below: The $RSSI_{type}$ value was set to -75 when $type$ is 'indoor' and -79 when $type$ is 'corridor'.

1) The type of the beacon location was obtained from the information in the movement log, based on a correspondence table.
2) If the RSSI in the beacon log was lower than that of the threshold $RSSI_{type}$, the information was considered unreliable and was removed from the movement log.
3) If the accuracy in the GPS log was lower than that of the threshold ACC_{type}, it was removed from the movement log.

B. Creating a Candidate Point Cloud Series

This section describes the method for converting the movement logs to candidate point cloud series using the combination and filtering processes. A candidate point set represents a set of nodes in the beacon network that can exist at time t, and a candidate point set series represents a series of data that arranges candidate point sets in chronological order.

Two types of movement logs are included: beacon and GPS. The candidate point cloud therefore includes two groups of candidate points, one generated from beacon logs and the other from GPS logs.

The following is the procedure for converting beacon logs to candidate point clouds: At time t, beacon nodes corresponding to the top k beacons with the highest RSSI form the candidate point group.

The following is the procedure for converting GPS logs to candidate point clouds: Beacon nodes are obtained within a circle of radius R centered on the latitude and longitude coordinates estimated using GPS at time t. Subsequently, the top k beacon nodes closest to the estimated location are selected as the candidate point group at that time.

Finally, the set of candidate points generated from both beacon and GPS logs are combined to determine the set of candidate points at time t.

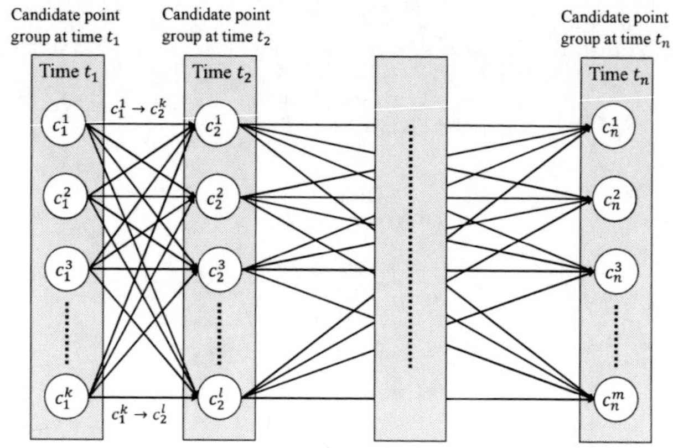

Fig. 6. Example candidate point graph. The graph is used for determining estimated path.

C. Estimating Routes From Candidate Point Cloud Series

In this section, we propose a method based on BST matching [2] for estimating a user's travel path from a candidate point cloud series and beacon network.

Figure 6 shows an overview of determining estimated path. The scores between each candidate point and candidate point at time t are computed. For each candidate point group, the path with the lowest overall score P is the path obtained by the Viterbi algorithm; this is the estimated path.

Score P is expressed as follows:

$$P = argmax\,(c_1, c_2, \cdots, c_n)\left\{ N\left(c_1^j\right) + \sum_{i=2}^{n} F\left(c_{i-1}^t \to c_i^s\right) \right\}$$

where $N\left(c_i^j\right)$ and $F\left(c_{i-1}^t \to c_i^s\right)$ represent functions that computes scores of candidate points and scores between candidate points, respectively. By calculating beacon-based and GPS-based candidate points together, the candidate point with the highest estimation accuracy can be selected for each point in time, regardless of the type. Candidate point score $N\left(c_i^j\right)$ represents the proximity between the observation point and candidate point c_i^j. The higher the proximity, the higher the score, and vice versa. The distance d_B between the beacon and observer can be obtained from the RSSI of the beacon received by the observer using the Friis transmission equation, defined as follows:

$$d_B = 10^{\frac{1}{20}(RSSI_{max} - RSSI)}$$

where $RSSI_{max}$ denotes the RSSI value when a radio wave is received at a distance of 1 m from the beacon. This value varies based on the location of the beacon; in this study, it is set at -55 for beacons in rooms and -60 for beacons in corridors.

TABLE III
NUMBER OF BEACONS INSTALLED IN THE EXPERIMENTAL ENVIRONMENT.

Location	Number of installed beacons
Bldg. 52, 1st floor	11
Bldg. 52, 2nd floor	32
Bldg. 23, 1st floor	25
Bldg. 23, 2nd floor	22

The beacon-based score $N\left(c_i^j\right)$ of a candidate point c_i^j is the reciprocal of distance d_B and is expressed as follows:

$$N\left(c_i^j\right) = \frac{1}{d_B}$$

The GPS-based score $N\left(c_i^k\right)$ of a candidate point c_i^k is the reciprocal of distance d_G between the estimated position at time t_i and the candidate point c_i^k, expressed as follows:

$$N\left(c_i^k\right) = \frac{1}{d_G}$$

Intra-candidate score $F_t\left(c_{i-1}^t \to c_i^s\right)$ indicates the travel speed between time t_{i-1} to t_i on the estimated path at average walking speed v_a. This value approaches zero as average walking speed increases.

$$F_t\left(c_{i-1}^t \to c_i^s\right) = \begin{cases} 1 & (v \leq v_a) \\ \dfrac{v_a t}{dist} & (v > v_a) \end{cases}$$

where $dist$ denotes the distance between two points on the network and t denotes elapsed time. v_a was set at $2.0\,\mathrm{m/s}$ for this study. This suppresses estimation of routes that would be considered unnatural at walking pace.

V. EVALUATIVE EXPERIMENT

In this section, we conducted an experiment to evaluate the accuracy of route estimation and generation time for validating the efficacy of the proposed method.

A. Experiment Conditions

In this experiment, participants were asked to carry a smartphone and walk along a designated route to verify the accuracy of the actual route and estimated route. The experiment was conducted in Buildings 52 and 23 of the Nagoya Institute of Technology. The Buildings 52 and 23 have 3 and 4 floors, respectively. Table III lists the total number of beacons installed in each floor. The routes used in the experiment spanned indoor and outdoor areas: three routes were set up, as shown in Figure 7. Movement logs were collected 6 times for each route.

The following five methods of comparison were used:

m1 (Proposed method): Path estimation based on global map matching using beacon and GPS

m2 Route estimation that prioritizes beacons with ancillary use of GPS

m3 Route estimation that prioritizes GPS with ancillary use of beacons

m4 Route estimation using beacons only

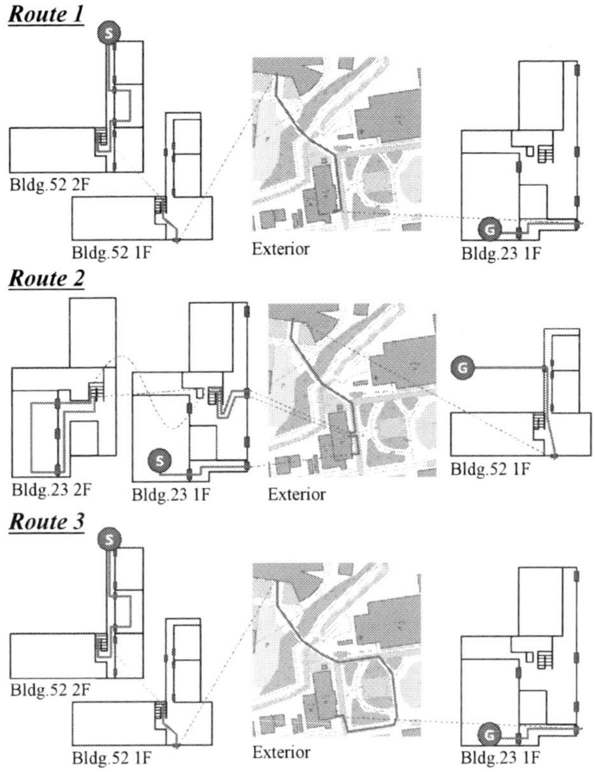

Fig. 7. Three routes used in the experiment.

m5 Route estimation using GPS only

m1 is our proposed method, which optimally uses GPS and beacons together. By contrast, **m2** and **m3** use both beacon and GPS data, however give preference to one type over the other. **m4** and **m5** use only one of the data types.

Precision, Recall, and F-score for estimation accuracy and generation time were used as evaluation metrics. Considering V_1 as the set of links in the routing network estimated from movement logs and V_2 as the set of links actually traversed, Precision P_V, Recall R_V, and F-score F_V are obtained as follows:

$$P_V = \frac{|V_1 \cap V_2|}{|V_1|}, \quad R_V = \frac{|V_1 \cap V_2|}{|V_2|}, \quad F_V = \frac{2}{\left(\frac{1}{P_V} + \frac{1}{R_V}\right)}$$

B. Experiment Results

We conducted an experiment and got results. Figure 8 shows the averages of the experiment results. The proposed method, **m1**, with an average F-score of 0.87, outperformed the other methods (0.84 for **m2**, 0.83 for **m3**, 0.76 for **m4**, and 0.67 for **m5**). The proposed method, **m1**, also yielded the best results for precision and recall.

m1, **m2**, and **m3** are methods that use both GPS and beacon data, however **m1** yielded the best results. This suggests that, compared to the methods prioritizing beacons (**m2**) and GPS

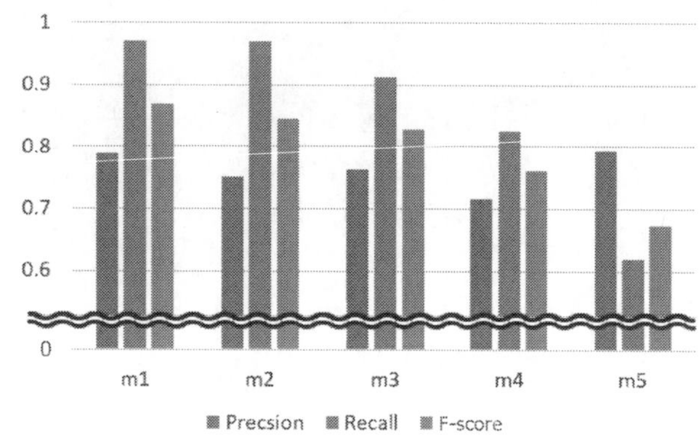

Fig. 8. Average results from the experiment.

TABLE IV
RESULTS OF A T-TEST. COMPARISON OF EACH METHOD AND **M1**.

	Route 1	Route 2	Route 3
m2	0.007	0.051	0.019
m3	0.054	0.009	0.092
m4	0.007	0.003	0.002
m5	0.000	0.012	0.000

(**m3**), it is more effective to use data based on the evaluation equation, as in our proposed method.

A t-test was conducted for each route to observe if there was a significant difference between **m1** and the other methods. Table IV lists the results of these t-tests. In all cases, the results were below the 0.1 significance level, supporting the superiority of the proposed method **m1**. The significance level was not below 0.05 for several items. This can be attributed to the fact that only six movement logs were collected. In future, we would like to increase the number of experiments, repeating the experiment to show significance.

Tables V and VI list the execution time results and the number of nodes at that time, respectively. Table V shows that **m4**, which uses only beacons for estimation, took the shortest computation time, while **m1** took the longest. Note that **m4** also has the smallest number of nodes in the candidate point graph used for estimation of all methods, while **m1** has the highest number of nodes, as shown in Table VI. Thus, it can be concluded that the computation time required was based on the number of nodes used for estimation.

Of the five estimation methods, **m1**, **m3**, and **m5**, which required longer computation timed, are likely to generate large numbers of candidate points using GPS data. The number of nodes in the candidate point graph therefore increased compared to **m2** and **m4**, which have shorter computation times, resulting in longer computation times.

VI. CONCLUSION AND FUTURE WORK

In this study, we proposed a route estimation method based on global map-matching that uses both BLE beacons and GPS. By using BLE beacons with GPS, it is possible to estimate routes that cross indoor and outdoor areas. We also

TABLE V
COMPUTATIONAL TIME TAKEN TO ESTIMATE.

[ms]

	m1	m2	m3	m4	m5
Route 1	13575	5993	12205	1164	11701
Route 2	24889	8326	19315	1018	23903
Route 3	23655	14663	20383	841	20423

TABLE VI
NUMBER OF NODES IN THE CANDIDATE POINT GRAPH USED FOR ESTIMATION.

	m1	m2	m3	m4	m5
Route 1	508	331	469	110	398
Route 2	937	476	848	152	785
Route 3	803	607	757	92	671

implemented a prototype system for achieving this feature and conducted an evaluative experiment using the prototype system. The results of the experiment indicated that using the BLE beacon and GPS-based scores produced better results. In the future, we will attempt to increase the accuracy of the method. In addition, because the evaluation in this study was conducted using only a small amount of data, we want to conduct a larger-scale demonstration experiment in future.

REFERENCES

[1] S. Kajioka, T. Mori, T. Uchiya, I. Takumi, and H. Matsuo, "Experiment of indoor position presumption based on RSSI of bluetooth LE beacon," in *Proceedings of the IEEE 3rd Global Conference on Consumer Electronics (GCCE 2014)*, pp. 337–339, 2014.

[2] D. Yamamoto, R. Tanaka, S. Kajioka, H. Matsuo, and N. Takahashi, "Global map matching using BLE beacons for indoor route and stay estimation," in *Proceedings of the 26th ACM SIGSPATIAL*, pp. 309–318, 2018.

[3] S. Brakatsoulas, D. Pfoser, R. Salas, and C. Wenk, "On map-matching vehicle tracking data," in *Proceedings of VLDB*, pp. 853–864, 2005.

[4] Y. Lou, C. Zhang, Y. Zheng, X. Xie, W. Wang, and Y. Huang, "Map-matching for low-sampling-rate gps trajectories," in *Proceedings of SIGSPATIAL GIS 2009*, pp. 352–361, 2009.

[5] P. Newson and J. Krumm, "Hidden Markov map matching through noise and sparseness," in *Proceedings of the ACM SIGSPATIAL 2009*, pp. 336–343, 2009.

[6] K. Urano, K. Hiroi, K. Kaji, and N. Kawaguchi, "A location estimation method using BLE tags distributed among participants of a large-scale exhibition," in *Adjunct Proceedings of the 13th International Conference on Mobile and Ubiquitous Systems: Computing Networking and Services*, pp. 124–129, 2016.

[7] M. Ficco, F. Palmieri, and A. Castiglione, "Hybrid indoor and outdoor location services for new generation mobile terminals," *Personal and Ubiquitous Computing*, vol. 18, pp. 271–285, 2014.

[8] S. Ito and N. Kawaguchi, "Wireless LAN based hybrid positioning system using bayesian inference and access point selection," *Transactions of the Institute of Electrical Engineers of Japan. C*, vol. 126, no. 10, pp. 1212–1220, 2006.

[9] C. Xiao, D. Yang, Z. Chen, and G. Tan, "3-D BLE indoor localization based on denoising autoencoder," *IEEE Access*, vol. 5, pp. 12751–12760, 2017.

[10] K. Urano, K. Hiroi, T. Yonezawa, and N. Kawaguchi, "An end-to-end BLE indoor localization method using LSTM," *Journal of Information Processing*, vol. 29, pp. 58–69, 2021.

[11] S. Kajioka, T. Saito, D. Yamamoto, and N. Takahashi, "A web-based geospatial adjacency graph construction tool for indoor/outdoor beacons," in *Proceedings of the IEEE 10th Global Conference on Consumer Electronics (GCCE 2021)*, pp. 1–2, 2021.

Autonomous Drone Landing in 3D Urban Environment Using Real-Time Visibility Analysis

Oren Gal and Yerach Doytsher
Mapping and Geo-information Engineering
Technion - Israel Institute of Technology
Haifa, Israel
e-mails: {orengal,doytsher}@technion.ac.il

Abstract— **Quadcopters are four rotor Vertical Take-Off and Landing (VTOL) Unmanned Aerial Vehicle (UAV) with agile manoeuvring ability, small form factor and light weight – which makes it possible to carry on small platforms. Quadcopters are also used in urban environment for similar reasons – especially the ability to carry on small payloads, instead of using helicopters on larger vehicle which are not possible in these dense places. In this paper, we present a new approach for autonomous landing a quadcopter in 3D urban environment, where the first stage is based on free obstacle environment and maximal visibility for the drone in the palled landing spot. Our approach is based on computer-vision algorithms using markers identification as input for the decision by Stochastic Gradient Descent (SGD) classifier with Neural Network decision making module. We use OpenCV with its built-in ArUco module to analyse the camera images and recognize platform/markers, then we use Sci-Kit Learn implementation of SGD classifier to predict landing optimum angle and compare results to manually decide by simple calculations. Our research includes real-time experiments using Parrot Bebop2 quadcopter and the Parrot Sphinx Simulator.**

Keywords-Swarm; Visibility; 3D; Urban environment; autonomous landing.

I. Introduction and Related Work

A Quadcopter is a specific type of a UAV, with four rotors and Vertical takeoff and Landing (VTOL) capability, its agility, light weight and size makes it a perfect companion to smaller boats from sail-boats to even kayak, rather than classic helicopters that accompany bigger ships or fixed-wings airplanes on extremely large aircraft carriers.

In the many uses of UAV (Unmanned Aerial Vehicle) a pilot uses real-time telemetry to take-off, fly and land the craft with continuous communication between ground station and the UAV on-board computer. Making these tasks Autonomous, will allow UAVs to perform missions without continuous communication, and thus prevent hijack or damage by hackers, be more stealth for surveillance and have unlimited distance from GS (bound to energy limitation).

Autonomous landing of a UAV is a problem on the focus of many studies [5] [6] [7] and landing on marine vessel makes this problem even more complex due to sea level motion that also occur when target platform is at stand-still.

The object of this research is to produce a safe landing mechanism for a quadcopter in 3D urban environment, in order to allow it to perform fully autonomous missions carried out at sea. Also, this mechanism could be used in pilot guided missions, as guideline suggestions to the pilot with how/when it is safe to land.

We assume the target position is known and Ground Station sets "home" position in the drone to be target's GPS position. Then the Bebop2 built-in "Return Home" function will bring it to the target, with up to a few meters off.

The proposed mechanism will perform the following tasks to achieve a "safe landing" decision: First, we need to visually search for and recognize the platform target and find the docking area. Once the target is found, the drone should set course and fly to target to be exactly above. Then, we detect and analyze the position of the landing surface and its plane angle relative to the camera. And finally, we will send the data to each of two implementations of the decision algorithms: 1. Using a supervised machine-learning classifier (pre-loaded with data), The machine input requires a quick pre-processing to set the data into a fixed structure vector, to resemble fitted data in the classifier. 2. Calculating directly from the data returned from the ArUco detection functions. The drone will then land safely on the boat, by sending a "land" command on time.

The problem of autonomous landing an UAV was on the focus of many studies as the survey review state-of-the art methods of vision-based autonomous landing, for a wide range of UAV classes from fixed-wing to multi-rotors and from large-scale aircrafts to miniatures. The main motivation for dealing with autonomous landing is the difficulty in performing a successful landing even with a pilot controlling the UAV. As it seems by statistics showed in [4], most of the accidents related to Remotely Piloted Aircraft Systems (RPAS) occur when the pilot tries to land the UAV.

Extensive research has been done on the subject to explore the various situations, technologies and methods to engage this problem. The work performed on previous studies, reviewed later in this section, is a great starting point for this project, as it is purely academic and relays on series of already existent technologies and tools, such as OpenCV [3], Sci-kit learn and the Parrot Ground SDK [1].

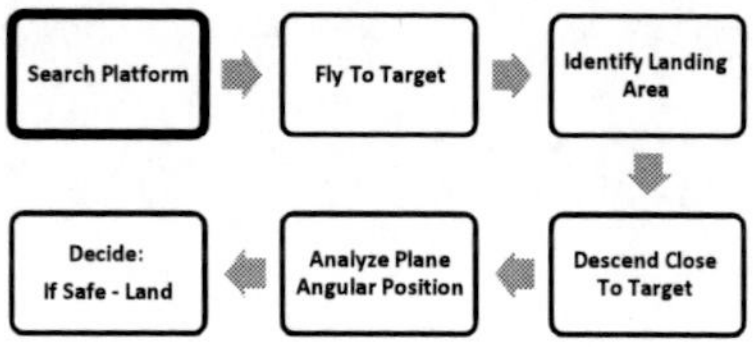

Figure 1: Proposed autonomous landing mechanism

In the following sections, we first introduce an overview of 3D models and extended the 3D visible volumes analysis. In the next section, we present the autonomous navigation process based on our fast visibility analysis with training data and classifier as can be seen in Figure 1. Later on, we present the simulation based on our 3D visible volumes analysis.

II. AUTONOMOUS NAVIGATION PROCESS

The basic step starting this process related to obstacle avoidance and visible area described in the next sections. Following that, we divide the autonomous navigation mission into two separate problems. The first part deals with navigating UAV from an arbitrary position far from target, as far field. Second part related to navigating to the target in the near field where the target is visible.

In the first scenario, which is when the mission objectives are reached and the drone needs to get to the target vessel for landing, we can use the built-in functionality of the drone to "Return Home" by setting it "Home" position to the target's known GPS position.

Bebop2 "Return Home" function works in a way that it will lift the drone to 20m above ground relative to take-off position, then fly directly to GPS position of "Home" and descend to 2m. Notice that if the drone is starting at height of more than 20m it will not descend to 20m, but rather keep its height until final descend near "Home".

The "Return Home" accuracy brings the drone to "Home" sometimes with offset of a few meters. This is good enough to get us to the second problem of navigation with visual distance to the target, until the drone will be directly above target and ready for landing.

Once the drone is at "Home" position, it will rotate and with each full rotation the tilt angle will increase to look further below, and if after rotating and tilting to the maximum of -90 degrees to the horizon, i.e. directly down, it will try again at higher altitude (1m up) to maybe see further away.

After getting a visual identification the drone will set course, keeping the target in the middle of the screen, and moving forward to it, tilting the camera during the movements until the landing pad is directly below. According to that, landing pad located in the middle of the image and camera tilt is maximum.

Then the drone will lower altitude to ~50cm while keeping the landing pad centered underneath, and in that height the data from the AR tags will be converted to a vector of predefined structure to feed a classifier trained to detect optimum landing angle/position. Once the classifier gives "Safe" signal – a "Land" command will issue to the drone to perform immediately.

III. FAST AND APPROXIMATED VISIBILITY ANALYSIS

In this section, we present an analytic analysis of the visibility boundaries of planes, cylinders and spheres for the predicted scene presented in the previous sub-section, which leads to an approximated visibility. For the plane surface, fast and efficient visibility analysis was already presented in [29].

In this part, we extend the previous visibility analysis concept [29] and include cylinders as continuous curves parameterization $C_{c\ln d}(x,y,z)$.

Cylinder parameterization can be described in (1):

$$C_{C\ln d}(x,y,z)=\begin{pmatrix} r\sin(\theta) \\ r\cos(\theta) \\ c \end{pmatrix}_{r=const}, \quad \begin{matrix} 0\le\theta\le 2\pi \\ c=c+1 \\ 0\le c\le h_{peds\_\max} \end{matrix} \tag{1}$$

We define the visibility problem in a 3D environment for more complex objects as:

$$C'(x,y)_{z_{const}}\times(C(x,y)_{z_{const}}-V(x_0,y_0,z_0))=0 \tag{2}$$

where 3D model parameterization is $C(x,y)_{z=const}$, and the viewpoint is given as $V(x_0,y_0,z_0)$. Extending the 3D cubic parameterization, we also consider the case of the cylinder. Integrating (1) to (2) yields:

$$\begin{pmatrix} r\cos\theta \\ -r\sin\theta \\ 0 \end{pmatrix}\times\begin{pmatrix} r\sin\theta-V_x \\ r\cos\theta-V_y \\ c-V_z \end{pmatrix}=0 \tag{3}$$

$$\theta=\arctan\left(-\frac{-r-\frac{\left(-vy\,r+\sqrt{vx^4-vx^2r^2+vy^2vx^2}\right)vy}{vx^2+vy^2}}{vx},\; -\frac{-vy\,r+\sqrt{vx^4-vx^2r^2+vy^2vx^2}}{vx^2+vy^2}\right) \tag{4}$$

As can be noted, these equations are not related to Z axis, and the visibility boundary points are the same for each *x-y* cylinder profile, as seen in (3), (4).

The visibility statement leads to complex equation, which does not appear to be a simple computational task. This equation can be efficiently solved by finding where the equation changes its sign and crosses zero value; we used analytic solution to speed up computation time and to avoid numeric approximations. We generate two values of θ generating two silhouette points in a very short time computation. Based on an analytic solution to the cylinder case, a fast and exact analytic solution can be found for the visibility problem from a viewpoint.

We define the solution presented in (4) as x-y-z coordinates values for the cylinder case as Cylinder Boundary Points (CBP). CBP, defined in (5), are the set of visible silhouette points for a 3D cylinder, as presented in Figure 2:

$$CBP_{i=1..N_{PBP\_bound}=2}(x_0, y_0, z_0) = \begin{bmatrix} x_1, y_1, z_1 \\ x_{N_{PBP\_bound}}, y_{N_{PBP\_bound}}, z_{N_{PBP\_bound}} \end{bmatrix} \quad (5)$$

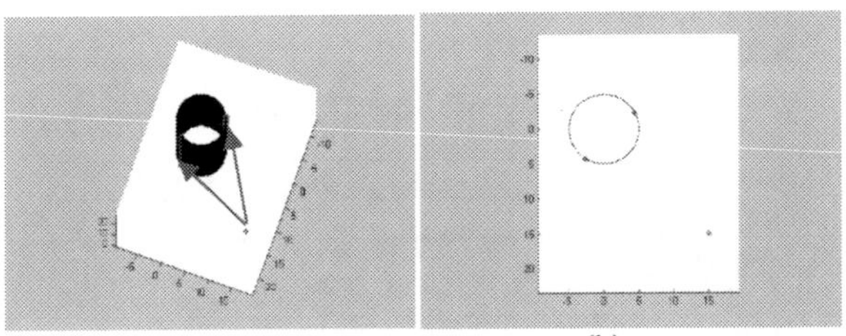

(a) (b)

Figure 2: Cylinder Boundary Points (CBP) using Analytic Solution marked as blue points, Viewpoint Marked in Red: (a) 3D View (Visible Boundaries Marked with Red Arrows); (b) Topside View.

In the same way, sphere parameterization can be described as formulated in (6):

$$C_{Sphere}(x, y, z) = \begin{pmatrix} r\sin\phi\cos\theta \\ r\sin\phi\sin\theta \\ r\cos\phi \end{pmatrix}_{r=const} \quad (6)$$
$$0 \le \phi < \pi$$
$$0 \le \theta < 2\pi$$

We define the visibility problem in a 3D environment for this object in (7):

$$C'(x, y, z) \times (C(x, y, z) - V(x_0, y_0, z_0)) = 0 \quad (7)$$

where the 3D model parameterization is $C(x, y, z)$, and the viewpoint is given as $V(x_0, y_0, z_0)$. Integrating (6) to (7) yields:

$$\theta = \arctan\left(\frac{r\sin(\phi)}{v\_y} - \frac{1}{v\_y\,(v\_y^2 + v\_x^2)}\left(v\_x\left(r\sin(\phi)\,v\_x - \sqrt{-v\_y^2 r^2 \sin(\phi)^2 + v\_y^4 + v\_x^2 v\_y^2}\right)\right), \frac{r\sin(\phi)\,v\_x - \sqrt{-v\_y^2 r^2 \sin(\phi)^2 + v\_y^4 + v\_x^2 v\_y^2}}{v\_y^2 + v\_x^2}\right) \quad (8)$$

Where r is defined from sphere parameter, and $V(x_0, y_0, z_0)$ arechanges from visibility point along Z axis, as described in (8). The visibility boundary points for a sphere, together with the analytic solutions for planes and cylinders, allow us to compute fast and efficient visibility in a predicted scene from local point cloud data, which are updated in the next state.

This extended visibility analysis concept, integrated with a well-known predicted filter and extraction method, can be implemented in real time applications with point clouds data.

IV. VISIBILITY-BASED DRONE AUTONOMOUS LANDING

The landing pad designed as a plate with five markers – one in the center and four others on each corner:

Figure 3: Landing pad with fiducial markers

Every ArUco marker has an ID as described in Figure 3, which can be determined when the marker gets detected, and by that we can easily center the drone location above the landing pad even if only one or two markers are in view.

This landing pad has markers with ID values of: [18, 28, 17, 25, 4] selected randomly, but once selected they are very important to the implementation since the training data linked to the classifiers used as will be discussed later.

The proposed system takes each frame, and resolve all markers, then create a data vector of fixed length with all the necessary information of the markers.

Data format for each marker can be described as: [ID, rx, ry, rz, tx, ty, tz],

Where $\begin{bmatrix} r_x \\ r_y \\ r_z \end{bmatrix}$ is the rotation vector of a single marker and $\begin{bmatrix} t_x \\ t_y \\ t_z \end{bmatrix}$ is the translation vector of that marker. This format repeats five times in each vector, where a tag ID has a fixed position for each tag. When a marker could not be found on a frame, the tag ID and all values of that marker will be set to zero. Then send this vector to a classifier which will simply return strings telling us if the drone is centered above the landing pad or a correction movement is required. Possible answers are in the set: "CENTER", "DOWNWARD", "FORWARD", "RIGHT", "LEFT".

For the Navigation we added more ArUCO tags surrounding this pad, in three sizes, so that they will be visible from varying distances along the navigation and descend process of the mechanism.

We used eight large tags, each surrounded by four medium tags and in between another five small tags as seen in Figure 4. The landing pad is printed on A4 page. And each of the eight patterns described here is also on an A4 pages.

Figure 4: Navigation Assisting Tag Board Design

A. Training Data and Classifiers

In order to train the classifier, we used OpenGL as can be seen in Figure 5 to simulate the landing pad in a precisely controlled position and viewing angles. By that, we created a labeled data set, then use this precisely labeled data to fit in a variety of classifiers and test for accuracy. Following that, we tested several classifiers and selected the best performance for the purpose of the landing mechanism proposed.

Figure 5 - OpenGL Simulation for Training Data

The simulated platform rotating in roll, pitch and yaw - controlled by passing parameters, allowing me to tag every rendered frame as either safe for landing or not without visual computation (pre-label the data).

The simulator gets parameters from command line for setting some axis angle to run on a limited range, while rolling over all possible values of angles and positions, so the workload could be divided to parallel processes and even run on different machines.

After a few days running on several computers in parallel, the simulators generated a total of 15,193,091 vectors dataset, that could be used as training dataset for different models of classifiers.

Sci-Kit Learn package implements SVM with a fit function that takes labeled data as input in two variables: Y vector of y labels in a single column and X array of x vectors – each x vector is a line vector corresponds to the appropriate y in Y.

SVM does not allow incremental learning, i.e., it needs all data at once. This was quit an issue with the data size we tried to fit – fifteen million vectors. However, Sci-Kit Learn offers other types of classifiers, although all of them do not perform actual incremental learning (they do need all data at once), nonetheless, they do implement a partial fit function that can take each round a small portion of the data, and update the classifier's support vectors.

For each classifier, we tried different parameters, and different sizes of the dataset by selecting randomly a fraction of the data. Then, test the model (using 25% of the data for test) to check it prediction accuracy.

TABLE I. CLASSIFIERS ACCURACY COMPARISSION

Classifier Type	Best accuracy
SGD, epsilon insensitive	57.341%
SGD, hinge	75.716%
SGD, huber	59.841%
SGD, log	73.658%
SGD, modified huber	73.362%
SGD, squared eps. insensitive	59.6%
SGD, squared hinge	73.857%
SGD, squared loss	57.171%
Perceptron	74.579%
Bernoulli NB	62.317%
Passive Aggressive Classifier	74.455%

The result in Table I show that even the best classifier got only approximately 75% success in recall. This is insufficient for a safety mechanism even with filters added to the process of a final "safe" decision.

To further increase accuracy, we thought it would be more effective to use more than one classifier, in a voting manner, to decide together on the data. At first, we suggested a voting scheme that takes 10-15 of the best classifiers and check if more than 50% of them agree on a "safe" result, take that as the answer, we checked that over the data and results did not increase accuracy at all. Then we thought maybe a classifier of classifiers outputs could extract some new information in a smarter manner than a simple voting, and will help increase accuracy. We created a new dataset of the same size, only this time the vector consisted of zero for safe and one for unsafe result of a classifier over fifteen of the best classifiers (72%-75% accuracy) and trained this dataset on all types of classifiers with different parameters as before. This time, all classifiers listed above got around 76% accuracy, where the best classifier reached 76.8% accuracy. Approximately 2% improvement.

Finally, looking closely on live videos of the ArUco markers detections, we noticed that the axis drawn on the detected markers tend to shift rapidly usually around more "safe" angles, so we tried to manually correct the data, and remove some of the spiking data that is tagged as safe – i.e. the simulator created it as a safe angle, but detection errors made it as a vector that should rather be tagged as unsafe.

All data marked as safe, with "Z" axis angle in all detected markers, re-tag as unsafe, if a certain threshold is passed.

Before rectifying the dataset consisted of about 50% safe labels. This method reduced the number of "safe" tagged vector to about 20% of the data.

Fitting this new retagged dataset to all models as before, and testing again for accuracy, results improvements shown in details reported in Table II. The results improved drastically.

Best classifier selected for the mechanism is SGD (Stochastic Gradient Descend) with loss parameter set to logarithmic. This classifier showed 86% percent accuracy, which could be used with some filtering to suppress false alarm rate even more.

V. SIMULATIONS

The quadcopter we used in this research is a Parrot Bebop2 drone. It is a GPS drone with full HD 1080p wide-angle video camera with 3-axis digital stabilization, that can also take 14MB still pictures.

Bebop2 has GPS guided Return Home feature, strong 6" propellers, long rage communication (with WiFi extender or Skycontroller remote), which makes it suitable for a windy outdoors flight.

The Bebop2 drone uses seven different sensors simultaneously to keep it stable and produce an extremely stabilized video even when the drone makes tiny maneuvers to keep itself in place, the apparent view to the user looks like the drone is in fixed position as if it was hanging on a crane. Also, there are no moving parts when we pan/tilt the camera, it is done entirely by changing the relevant pane in the full fisheye image.

Parrot Ground SDK includes software development suite that provides a tool for developers to communicate and control with Parrot drones that uses AR.SDK3 framework, e.g., Mambo, Bebop, Disco, and Anafi. It also includes a simulator platform called Sphinx, built on Gazebo platform, with Parrot drones not just as models but with full featured firmwares that are similar to the ones on the equivalent physical drones. This allows developers to fully test and debug their programs with real firmware feedback from a drone in mid-flight without the risk of injury or damages to equipment.

Ground SDK also provides a python wrapper called Olympe, to easily control drone objects. We preferred a third-party implementation named pyparrot, which is better documented and fully open-sourced, so it would be easier to add or change functionality to my needs.

A. *ArUco Markers*

The first problem we had to deal with, involves detection and identification of the landing pad. Afterword, we had to gather all planar information to pass to the decision mechanism for processing.

In order to simplify detection and get a fast and robust identification and planar information of the target, we used AR-tags on a specially designed landing pad. Specifically, the use of off-the-shelf open source ArUCO seem to be a simple solution (other implementations of AR-tags e.g. APRIL-TAGS may be suitable as well).

Implementation of ArUco marker detection exists in open-source library OpenCV, available for c/c++ and python. In order to get the marker real-world coordinates, we need the projection matrix of the camera and the distortion coefficients vector. To get these parameters a calibration is needed to be done once, then it could be loaded through a configuration file. The calibration process also available in OpenCV documentation, using a printed checkboard of known dimensions, and about twenty shots in different orientations and locations across the screen.

We incorporate different marker sizes to be able to detect markers in different distances from the target landing pad and follow the tags. ArUco Markers also have tag ID encoded in them so we even know which tag we are seeing and thus what size it is or where it is located on the board.

TABLE II. IMPROVEMENTS IN ACCURACY OF CLASSIFIERS

Classifier type	Best accuracy	Before correction	Improve ment
SGD, epsilon insensitive	83.450%	57.341%	26.11%
SGD, hinge	85.062%	75.716%	9.35%
SGD, huber	81.876%	59.841%	22.04%
SGD, log	**86.175%**	**73.658%**	**12.52%**
SGD, modified huber	86.131%	73.362%	12.77%
SGD, squared eps. Insensitive	82.019%	59.6%	22.42%
SGD, squared hinge	85.891%	73.857%	12.03%
SGD, squared loss	82.942%	57.171%	25.77%
Perceptron	85.470%	74.579%	10.89%
Bernoulli NB	81.664%	62.317%	19.35%
Passive Aggressive Classifier	84.041%	74.455%	9.59%

B. *Implementation*

To get control over a Bebop2 Drone, we found two python wrappers that we could use, and tested both of them. The first one comes with a Parrot Ground-SDK suite which includes the Sphinx Simulator, called Olympe. The Second wrapper pyparrot, originally developed for the Parrot Mambo, but now capable of controlling most of the newer generation Parrot drones.

We decided to work with pyparrot due to two main reasons: 1. Olympe used a closed virtual environment that made it harder to install additional packages using pip. 2. pyparrot is an open source, making it easy to adapt and change to my needs, it also suggests two types of video handling class: the first one uses FFMPEG and the other opens SDP file with VLC on a separate thread. Both methods were slow and missed critical frames especially in SEARCH mode, when the camera rotates to find the target. Sometimes the video smeared so badly we could barely recognize the landing pad even when we knew where it was there.

We changed the video handler to run on a separate thread (like the VLC option on pyparrot) only that in my implementation we used standard OpenCV capturing module VideoCapture to open SDP file (contains IP, port, codec) for streaming coming from the drone or sphinx (depends on DRONE_IP parameter in the code), and another separate thread for the automation state machine that runs the different stages of this autonomous mission control and landing mechanism.

For proof of concept, all experiments were simulated in Gazebo based Sphinx simulator without moving wave simulations, or any automated changes in landing-pad angles or position. The changes were made manually by rotating the pad during simulation when the drone was waiting to get a safe signal from either classifier or calculations.

The experiment also did not simulate the use of "Return Home" functionality and assumed to start near target at about five meters in a random position. The drone starts to search around to get a visual of the landing pad, then fly to set exactly above while looking directly down (-90 degrees below horizon). Drone initiate slow descend while keeping target still in the middle of the frame, until reaches height of less than 50cm. In this stage, decision mechanism under test should trigger "safe" when ArUco markers of the pad will be in a position that is regarded flat enough to be considered as safe.

In a preliminary experiment, we found that the classifier that we trained, could not get to a "safe" decision even when the landing pad was flat without any movements. Same classifier was tested with images from web-cam input seems to work fine, this could be issue caused by miscalibration of the camera. These inaccuracies cause ArUco functions that heavily relay on camera calibration, to produce different range of data relative to what the classifier was trained with (data from an OpenGL graphics drawn landing pad). This method should be further explored in future work.

Simplified manual calculation that work directly on data from ArUco functions output, could also be easily recalibrated and adjustable to fit with data ranges of mis-calibrated data. Finally, running full scenario of the experiment with landing pad on unsafe initial position got the drone flying above it and waiting, then manually flatten the landing pad, made the decision mechanism to trigger "safe" and send a landing command to the drone, which landed in the desired spot.

VI. CONCLUSION AND FUTURE WORK

In this work we introduced a mechanism for autonomous landing a quadcopter in. The work focused to assist in the final stage of an autonomous mission, when drone returned to home, but still needs to find exact position of landing on the target and dealing with sea-level motion of the target.

In this study we developed a training simulator to create large data set of visual input, produced by OpenGL graphics in a controllable manner. Also, we compared different types of trained classifiers to find best match to our particular data, and competed best classifier vs. direct observation

In conclusion, the ArUco functions produce enough information regarding marker positions to be used manually and get a satisfying result for that manner. It is fast and robust and easily read to get a quick answer to whether it is safe or not, and the use of a classifier is not necessary.

REFERENCES

[1] Parrot Inc., "developer.parrot.com," 2020. [Online]. Available: https://developer.parrot.com/docs/olympe/userguide.html.

[2] A. McGovern, "pyparrot github repository," JAN 2020. [Online] https://github.com/amymcgovern/pyparrot.

[3] OpenCV, Open Source Computer Vision Library, 2015.

[4] K. Williams, "A Summary of Unmanned Aircraft Accident/Incident Data: Human Factors Implications," The Federal Aviation Admistrator Oklahoma City, 2004.

[5] A. F. Cobo and F. C. Benıtez, "Approach for Autonomous Landing on Moving Platforms based on computer vision," The International Journal of Computer Vision, vol 4, 2016.

[6] L. Daewon, R. Tyler and K. H. Jin, "Autonomous landing of a VTOL UAV on a moving platform using image-based visual servoing," IEEE International Conference on Robotics and Automation, pp. 971-976, 2012.

[7] T. Merz, S. Duranti and G. Conte, "Autonomous Landing Of An Unmanned Helicopter Based On Vision And Inertial Sensing," Experimental Robotics IX, pp. 343--352, 2006.

Identifying and Visualizing Older Single-Family House Areas for a Sustainable Spatial Planning

Dominik Visca, Max Hoppe, Kevin Kaminski, Pascal Neis
Department of Technology
Mainz University of Applied Sciences
Mainz, Germany
emails: dominik.visca@hs-mainz.de, max.hoppe@hs-mainz.de, kevin.kaminski@hs-mainz.de, pascal.neis@hs-mainz.de

Abstract— **The paper focuses on the challenge of sustainable spatial planning in areas with older single-family houses in Germany, particularly those constructed between the 1950s and late 1970s. Demographic changes, especially in rural areas, have resulted in vacancies in these building stocks, making the identification of these areas crucial for an effective spatial planning. The construction year of buildings can be used to identify single-family houses areas, however, obtaining fine-grained information from official sources is limited. This paper presents a method to identify building year classes through the combination of various sources and data modelling. The method offers a practicable approach for both research and spatial planning. Thus, this paper reflects trends in the field of Urban Geo and Spatial Planning issues, highlighting the value of information to be able to analyze social change and corresponding infrastructures on a small-scale level.**

Keywords-sustainable spatial planning; disaggregation; open data; demographic changes; single-family houses

I. INTRODUCTION

In Germany, areas with older single-family houses, particularly those built between the 1950s and late 1970s, pose a challenge for a sustainable spatial planning process. Especially in rural areas, demographic changes are increasingly resulting in vacancies in the building stock of single-family houses, as transitions from first-time owners to young families or new owners are often problematic. Identifying these areas is therefore crucial, in order to take spatial planning measures.

The year of construction of a building can be used to identify relevant single-family house areas. Information pertaining to the construction years of buildings can be obtained from the AAA model - Official Control Point Information System (AFIS, Amtliches Festpunktinformationssystem), Official Real Estate Cadastre Information System (ALKIS, Amtliches Liegenschaftskatasterinformationssystem) and the Official Topographic Cartographic Information System (ATKIS, Amtliche Topographisch-Kartographische Informationssystem), which constitutes the official geobasic information provided by the Surveying and Cadastral Administrations in Germany (Vermessungsämter in Deutschland). However, this information is rarely found in actual data records. In more recent research work, zoning plans that are already available through an INSPIRE Directive (Directive 2007/2/EC) have been used as a makeshift. Yet, an analysis of planning regulations and the date of legal enforceability is merely an indicator of the construction year of residential buildings in a rezoned municipality - in shrinking regions, where vacant buildings often occur, there is sometimes a large discrepancy in time between the legally effective date of a zoning plan and the actual year of construction of a building. Single-family house areas cannot be conclusively identified there. Through the combination of various heterogeneous sources, and disaggregation of data, it is possible to identify required building year classes.

In order to analyze transformation processes, and to respond to them with appropriate spatial planning measures, data is needed. Since certain data are not provided at all scales - in particular, below the municipal level - or are even freely available, methods are required to overcome this deficit. Accordingly, this paper focuses on small-scale information gaps and methodological considerations that address these issues. To this end, existing approaches are taken up in order to smartly link aspects of them, resulting in new datasets needed. Thus, this approach provides both, a relevant contribution to information gathering and data modelling that can be picked up and reflected upon by other researchers and, in principle, a practicable method for questions of spatial planning issues and urban geo.

Before the method and data bases are outlined in Section III, Section II introduces into the background and challenges regarding sustainable spatial planning processes in the context of demographic changes in single-family house areas. In Section IV results are discussed and evaluated. A large-scale survey of official building footprints and the years of construction therein over the past 25 years serves this purpose as well. The article ends in Section V, with a conclusion and an outlook regarding future potentials of the method and data presented.

II. BACKGROUND AND CHALLENGES

In Germany, it is projected that by 2050, the population of individuals over the age of 80 will reach approximately 10 million, as compared to 6.1 million in 2021. In the same year, it is expected that 30 percent of the population will be over 65 years [1][2][3]. This demographic development poses a significant societal challenge and further accentuates the importance of addressing situations of transition or generational change in older single-family house areas [4][5][6][7]. These neighborhoods were initially built with the intention of offering sufficient space for families with children, and in addition to serve as part of an individual retirement security. Subsequently, there are changes caused by the life cycle: With the transfer to a suitable accommodation for elderly people or, at the very latest, when the parents' generation dies, there is a change of generation [4][5][6][7]. Many of the older single-family properties do not fulfil the current demands in terms of living space, room sizes or technical equipment. For the first-time owners who are not willing to invest, this means both a loss of property value and a type of housing that is no longer adequate - as the living space, which is often too large, only rarely meets the needs of the tenants [4][5][6][7]. As one consequence, the transition from one generation to the next becomes more difficult, especially in rural-peripheral locations and/or in economically less powerful regions.

In order to respond with adapted urban planning and spatial planning measures, fine-grained data is needed, based on which dynamics at a sub-municipal level can be analyzed, as well as construction year data for relevant building stocks. For small-scale demographic analyses, georeferenced resident registration data is increasingly used in Germany but due to privacy regulations not within easy reach [8][9].

Also, the availability of data pertaining to the construction years of buildings is often limited. Through the process of disaggregation and enrichment of data, this limitation can be surmounted, as demonstrated, for example, by Weck-Ponten et al. [10]. In this context, results from the 2011 Census, zoning plans, aerial and satellite images are often used. For example, Eichhorn & Siedentop [11] refer, in the context of estimating inner developments in municipalities in North Rhine-Westphalia between 1979 and 2011, to the construction year information in the 2011 Census data which they used in combination with aerial images. For Muckel and Ogorek [12], zoning plans have the advantage that, in addition to the type of use (see for Germany § 30 Baugesetzbuch, BauGB), they provide information on the construction year of the buildings visible in the map section or in the scope of validity through a binding date of legal force. At the same time, zoning plans not only offer a legal force date, they also can be seen as a historical document. Iosifescu, Tsorlini, Hurni [13], as well as Chen et al. [14] point out that vectorizing content from historical maps with sufficient accuracy for basic spatial queries is possible. In this regard, deriving building footprints based on aerial images [15], historical satellite images [16], and other vector datasets such as OpenStreetMap [17] is an area that has been extensively researched in recent years. However, this information database is only partly usable for vectorizing zoning plans. Herrmann [18] and Kment [19] point out, that due to the small scale and the mostly incompletely available types of buildings, no predictions about future developments for residential areas of the single-family houses can be realized.

To determine and classify single-family house areas in small municipalities by construction year, the 2011 Census data serves as a basis from which the dynamics of the building stock can be derived [20]. To disaggregate the census data, zoning plans, that are spatially available according to INSPIRE guidelines, are used, which can contain information on building construction years.

III. METHOD AND PROCEDURE

A. *Data Base*

Open data from the 2011 Census [21] serve as the basis for the identification and subsequent visualization of single-family residential areas. Since the German Federal Statistics Act (Bundesstatistikgesetz, BStatG) was amended in 2013, much of the data collected for the 2011 Census is available in aggregated 100m grid cells [22]. One of the characteristics collected is the classified year of construction of each residential building included in a grid cell. Table 1 shows the available specifications of the attribute "year of construction" in the Census 2011.

The list shows that the classification of the year of construction has a high temporal resolution, especially from 1987 onwards. In contrast, the period from the 1950s to the 1970s is covered by only one class "1949 - 1978". This class allows a delimitation of buildings to younger and older buildings, but no delimitation by decades within this class. The absence of the year 1979 and the additional presence of the year 1949 is negligible due to the lack of a nationwide available alternative of spatially high-resolution building year data.

TABLE I. SPECIFICATIONS OF THE ATTRIBUTE „YEAR OF CONSTRUCTION" OF THE 2011 CENSUS [22].

Coding in data set	Feature description
1	older than 1919
2	1919 – 1948
3	1949 – 1978
4	1979 – 1986
5	1987 – 1990
6	1991 – 1995
7	1996 – 2000
8	2001 – 2004
9	2005 – 2008
10	2009 onwards

B. *Identification of Single-Family Houses*

Adapting patterns from Pajares, Muñoz Nieto, Meng, Wulfhorst [23], and Visca, Hoppe and Neis [24], the construction year classes of the census data are to be disaggregated to residential buildings to get a more accurate level than the 100m grid cells provided. In a first step, the identification of residential buildings is necessary. For this purpose, the building footprints from ALKIS are used. In order to pre-filter side buildings (e.g., garages or subsequent extensions), only those building polygons appear useful that contain at least one geo-referenced building address, also taken from ALKIS. Subsequently, a filtering for residential buildings takes place by a comparison with ALKIS usage types on the one hand, and with OpenStreetMap building polygons with an "amenity" tag on the other hand, which exclude a corresponding residential building usage. For this purpose, address and building data per census grid cell are extracted from OpenStreetMap (OSM) through the Overpass API of Olbricht [25] for blending the results with the official building footprints. This removes buildings used exclusively for commercial, industrial, or agricultural purposes from the dataset.

In order to disaggregate the tabular building year classes onto the correct building polygons, information from secondary data sets is necessary. Here, the legally binding zoning plans of the municipalities are crucial. Regardless of the specific scope of a zoning plan, the entire map section of its planning framework provides information on the presence of buildings at the time of its legal validity date or the specified date of the cadastre data used. The absence of a zoning plan also provides an indication concerning the building year class. A mandatory zoning plan as a planning instrument did not come into place until the Federal Building Act (Bundesbaugesetz, BBauG) was passed in 1960; developed, but not zoned, larger areas must therefore be older [26]. For the zoned areas, a systematic evaluation of the building stock visible in the zoning plans is necessary.

C. Vectorization of Zoning Plans

Usually, the zoning plans provided are only available as scanned plan documents in PDF format, and sometimes also as a georeferenced raster image tailored to the area of application. Depending on the age of the zoning plan, from which the data is derived, plans made available in geoportals have exportable vector layers. For more recent CAD-based plans, this allows the PDF to be imported and the georeferenced vector data to be accessed in a geographic information system. In case of scans of signed originals or an older zoning plan, this option is not available. Thus, it is necessary to vectorize the zoning plans in order to match the existence of the building polygons present in the latest ALKIS inventory with the historical building polygons from ALKIS, the Automatic Real Estate Map (ALK, Automatisierte Liegenschaftskarte) or the cadastral map via a spatial query. The requirement to have building polygons that are as accurate as possible and rectangular is not necessarily required for querying whether or not a building exists at two (or more) different moments in time. The zoning plans are vectorized and assigned a timestamp of the legal date respectively the ALKIS data status.

D. Comparison of Building Footprints with Zoning Plans

As a next step, centroids are then to be formed within each polygon from the building footprints with (partial) residential use filtered at the very beginning. In order to identify the earliest presence of a building in a zoning plan, and to attach the corresponding timestamp to the building polygon, it is iteratively checked in chronological order for each zoning plan if the centroids formed are present within a polygon of similar area, from a vectorized zoning plan. As long as the approximate geometry and location is preserved, it is not necessary to map the building footprints as closely as possible. By the end of all iterations, every building polygon has a date for its earliest occurrence, and its last "non-occurrence" in a zoning plan, or "no indication" if the building is not visible in any zoning plan map section. The resulting time period corresponds to the time window in which the building was built.

E. Disaggreagation of the Census Data

The timestamps of the building polygons are then aggregated at a parcel level, always using the oldest building per parcel. All adjacent parcels with the same timestamp are then to be aggregated to obtain larger areas of the same building year references. These areas serve as the primary distribution feature for disaggregating the census data.

Per census grid cell, the actual disaggregation of building year classes is performed. For each "year of construction" attribute value contained in a grid cell, a three-stage distribution is performed on all buildings within the grid cell from the filtered building polygon dataset. First, all year-of-construction classes are distributed to those buildings that are located in an area of equal year-of-construction references appropriate for the year-of-construction class. The most specific time frame is to be used (i.e., "younger than 1985 and older than 2002" is more specific than "older than 2003"). If the distribution is not final, the relative distance to a known historic structure is added as an indicator of a building's age. Typical points of interest (POIs) in terms of this are churches, historic town halls or the market place as an indicator for the historic center of a village. Reflecting the zoning of settlement expansions, younger buildings tend to be further away from the historic village center than older ones. If building year classes still cannot be assigned to any building, they must be randomly distributed among all other buildings in the grid cell.

F. Identification of Single-Family Houses

The identification of single-family residential areas further necessitates the differentiation of single-family and multi-family buildings. For this purpose, the building data extracted via the Overpass API are intersected with the building polygons according to the procedure of Visca, Hoppe and Neis [24] based on the attribute "building type (size)" or the floor area and number of stories. Analogous to the aggregation of the building year information of the zoning plans to areas of identical building year references, a stepwise aggregation of the single-family houses to parcels and then a union based on these building year classes towards single-family house areas of identical building year classes has to be performed.

IV. RESULTS & DISCUSSION

The town of Otterberg (Verbandsgemeinde Otterbach-Otterberg, Rhineland-Palatinate, Germany) served as a test site. The town's settlement area covers approximately 2.20 km^2 and, according to ALKIS, contains 3,369 official building footprints, from which side buildings and buildings not used for residential purposes were filtered to 1,756 building footprints. A manual sorting out of properties in the external area, among others, was not carried out. The generated data set represents the basis for the following analyses.

In the study area, 22 zoning plans are legally binding and published on the Internet (as of July 2022). The dates of legal effect range from 1987 to 2019. A total of 1,081 of the 1,756 buildings used for residential purposes are covered by at least one zoning plan. An additional 45 buildings are outside of a zoning plan application area, but are included in the visible map section of at least one zoning plan. Thus, information for subsequent disaggregation is obtained for 1,126 buildings.

The 22 zoning plans were adapted to the map section, georeferenced and vectorized. The most significant spatial deviations exist for older plans, since their data basis is the old cadastral map. Due to the partially poor state of the plans, georeferencing proved to be difficult. As a result of the changed partitioning of the parcels, the reference points for the current cadastre are difficult to find. Furthermore, the data basis of the cadastral map also deviates spatially from one another within a plan, which makes it necessary to rectify and rotate certain sections. Especially, the old zoning plans require a substantial amount of manual preparation. The actual vectorization is not satisfactory to the same extent, given the very heterogeneous design of the zoning plans. In particular, zoning plans containing many hatchings often generate individual polygons for each hatching gap, which have to be generalized again in a post-processing step.

Figure 1 shows the 605 grey building footprints with a year of construction younger than 1987 that do not qualify as older single-family residential areas. The 521 red building footprints could be in the period being looked for. Often, however, the year of construction references are very vague, as seen in Table 2.

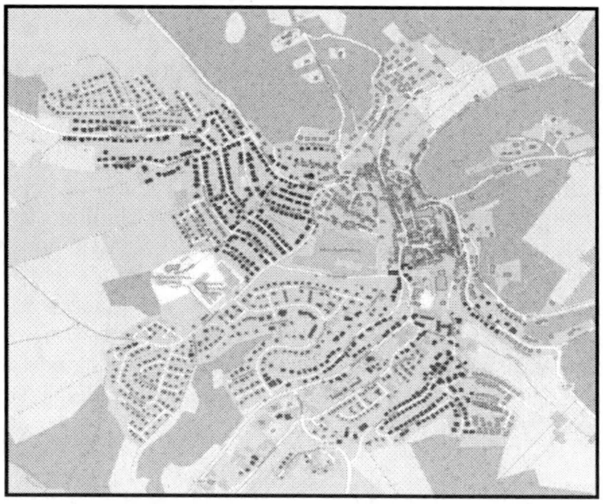

Figure 1. Years of construction derived from zoning plans, Map Tiles by Stamen Design, CC BY 3.0 Data from OpenStreetMap, ODbL.

The yellow building footprints are 630 buildings in the non-zoned town center whose construction years are older than 1960. These buildings could have been built in the early phase of settlement zoning after 1945. For the most part, these are buildings from the historic center of the village, which was built before 1919.

TABLE II. INDICATIONS OF CONSTRUCTION YEARS BASED OF ZONING PLANS.

Construction year information per zoning plan	Number of buildings
older than 2017	2
older than 2015	12
older than 2003	312
older than 2002	1
older than 2001	112
older than 2000	13
older than 1989	8
older than 1987	58

The disaggregation of the census data reveals that the residential areas of the 1950s to the 1970s have a high degree of coverage with the pre-analysis of the zoning plans due to the homogeneous settlement structure. Figure 2 shows a new zoning area that was planned and built upon after the cutoff date of the 2011 census. The white-colored building footprints in the right-hand map section imply that there is no information in the census data on a year of construction, while the grey building footsprints in the left-hand map section originate from the zoning plans and represent the years of construction after 1970. At the right, building blocks from 1949 to 1978 could be identified (red, right). Summations within the census data across several attributes are not always identical with the actual situation. The effects of the anonymization procedure can be seen at the top of the image: According to the zoning plans, these buildings are older than 2003 (red, left), but in the 2011 census there is no information (white, right) about them.

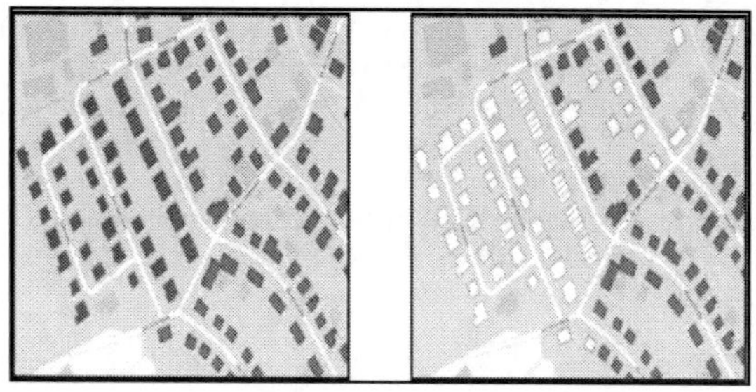

Figure 2. Comparison of zoning plan evaluation (left) and census disaggregation (right), Map Tiles by Stamen Design, CC BY 3.0 Data from OpenStreetMap, ODbL.

In the non-zoned, densely built-up inner area of the historic town center, the method produces only limited and unreliable results. For the sample grid cell, the census data set provides information on 35 buildings from six building year classes - see Figure 3. Due to the missing data from the zoning plan analysis and the similar distance between the properties and the village center, or POIs such as the church (bottom right of the image), only a random distribution is possible. Without further information, an allocation of building year classes by building or parcel is impossible. Nonetheless, in terms of the objective of defining single-family house areas, the map section shown can be ruled out as a possible match. Although it was possible to disaggregate a section of buildings on the left edge of the grid cell (see red area), it could hardly be considered an area of single-family houses. The yellow building footprints in the right-hand map section indicate the expected historic buildings from 1949 and earlier.

Figure 3. Example of insufficient disaggregation (left: zoning plan evaluation, right: census disaggregation), Map Tiles by Stamen Design, CC BY 3.0 Data from OpenStreetMap, ODbL.

The analysis clearly shows that the determination of the year of construction also requires the identification of each individual single-family house. A comparison by building type via OpenStreetMap holds little potential for success due to the tendency for data incompleteness in rural areas. A disaggregation of the building types from the census data set is therefore only possible randomised, whereby within the relatively homogeneous settlement expansion areas the aggregation level of the 100m zones can also provide sufficient accuracy. Neither necessary nor purposeful is the consideration of every single data outlier, whether of the year of construction or the building type. The single-family house areas are characterised by a mixture of the building stock through redensification or new buildings. Therefore, they must be considered in their entirety, regardless of some of the building stock that does not fit the definition. Figure 4 shows the aggregation of larger single-family house areas from the 1950s to the 1970s for the town of Otterberg.

Figure 4. Identified areas of older single-family houses, Map Tiles by Stamen Design, CC BY 3.0 Data from OpenStreetMap, ODbL.

To validate the results, a dataset of official building footprints from the State Office for Surveying and Geographic Information of Rhineland-Palatinate (LVermGeo, Landesamt für Vermessung und Geobasisinformation) is used, which supplements the construction years of the last approx. 25 years as part of a large-scale survey procedure. Although it cannot be used to identify the single-family housing areas, the information is useful as a cross-check to evaluate the methodology presented here. For 347 out of the 1,756 official building footprints, construction years between 1957 and 2021 could be added, with only 10 construction years older than 1998. About 44% of the building footprints were not taken into account by the disaggregation. For the most part, these are new buildings built after the 2011 Census cutoff date, which are not relevant to the focus of this research topic. For 19% of the remaining 56%, the disaggregation assigned a building year class that does not correspond to the actual building year. On the other hand, the method assigned the correct construction year class to the building footprints for approx. 81% of all relevant cases.

V. CONCLUSION AND OUTLOOK

With regard to future improvements of timeliness, depth, transferability and general availability of data, it can be concluded that updated building year classes will be available with the release of the 2022 Census results. This will allow a comparison with the 2011 Census results, revealing possible gaps caused by the anonymization process. In addition, a finer disaggregation could be performed to allow a more accurate identification of relevant single-family areas. For future demographic developments, precise information on current and prospective zoning is also obligatory.

Although the data of the real estate cadasters do have the advantage that they are subject to an ALKIS modeling regulation that is coordinated to all federal states, there are recognizable differences between the data products due to the decentralized management in more than 400 real estate offices. Furthermore, they are not available for free in all federal states [27][28]. A broader harmonization and an area-wide opening of the data sets would contribute to a better transferability of the demonstrated method. The data from the OpenStreetMap project benefits from continuous updating by contributors, which is also reflected in an increasing depth and availability of information in rural areas [29]. If incorrect or erroneous information is taken into account, an improved data reconciliation regarding land use types is potentially possible [30]. To obtain additional benefits for the analysis of visible building stock in zoning plans, access to plans that are no longer legally binding is of help. In reality, however, this is difficult to establish because the INSPIRE Directive (Directive 2007/2/EC) [31] only requires web-based access for zoning plans from December 2013 onwards. Historization of older zoning plans would be a desirable feature.

ACKNOWLEDGMENT

The work on this article was done as part of the project "Spatial Intelligence for the Integrated Care of Seniors in Rural Neighborhoods (RAFVINIERT)", which is funded by the Carl Zeiss Foundation in the program "Transfer - Intelligent Solutions for an Aging Society".

REFERENCES

[1] Statistical Offices of the Federal and State Governments. Population Projection. [Online] Available from: https://www.destatis.de/DE/Themen/Gesellschaft-Umwelt/Bevoelkerung/Bevoelkerungsvorausberechnung/_inhalt.html [retrieved: 02, 2023].

[2] Statistical Offices of the Federal and State Governments. Population status: Official population of Germany. [Online] Available from: https://www.destatis.de/DE/Themen/Gesellschaft-Umwelt/Bevoelkerung/Bevoelkerungsstand/_inhalt.html [retrieved: 02, 2023].

[3] E. Richter, Senior Democracy: The Aging of Society and Its Consequences for Politics, Berlin, Germany: Suhrkamp Verlag. 2020.

[4] B. Adam et al., "Older single-family house areas in transition - An underestimated planning challenge", Situation in North Rhine-Westphalia. ARL. 2018.

[5] T. Köhler and M. Schaffert, "Building Measures in the Face of Population Decline", in Challenges for governance structures in urban and regional development = Fragen zur Steuerung von Stadt- und Regionalentwicklung, pp. 195–208. Zürich, 2015.

[6] C. Krajewski, "What happens to dream houses? - Part 1: Dealing with single-family house settlements analyzed using the example of Altena and Finnentrop in Southwestphalia", in Institute of Geography, University of Münster, (14), Münster, 2014.

[7] A. Raab, "Demand-oriented inner development and land management - future issues in urban and rural redevelopment", in Mitteilungen DVW Bayern, 65(4), pp. 547-563, 2006.

[8] A. Lindner, "Municipal monitoring. KomMonitor: A digital planning tool for urban development", 47(3), pp. 76-87, 2020.

[9] M. Schonlau, C. Danowski-Buhren, M. Guth, U. Klein and, A. Lindner, "Integrated monitoring as a tool for sustainable urban development. IS THIS THE REAL WORLD? Perfect Smart Cities vs. Real Emotional Cities", in Proceedings of REAL CORP 2019, 24th International Conference on Urban Development, Regional Planning and Information Society, pp. 453-462, 2019.

[10] S. Weck-Ponten et al., Automated data aggregation for integrating a dynamic building simulation into a geographic information system. 2018.

[11] S. Eichhorn and S. Siedentop, "Inside before outside? An estimation of inner development in North Rhine-Westphalian municipalities, 1979 to 2011", Raumforschung Und Raumordnung | Spatial Research and Planning, 2022.

[12] S. Muckel and M. Ogorek, Public building law - outlines of the law. C.H. Beck, Munich, 2018.

[13] I. Iosifescu, A. Tsorlini and, L. Hurni, Towards a comprehensive methodology for automatic vectorization of raster historical maps. 2016. [Online]. Available from: https://www.e-perimetron.org/Vol_11_2/Iosifescu_et_al.pdf [retrieved: 02, 2023].

[14] Y. Chen, et al., "Vectorization of Historical Maps Using Deep Edge Filtering and Closed Shape Extraction", in Document Analysis and Recognition – Lecture Notes in Computer Science, Aufl. 12824. Springer, Cham, 2021.

[15] W. Li et al., "Semantic Segmentation-Based Building Footprint Extraction Using Very High-Resolution Satellite Images and Multi-Source GIS Data", Remote Sens, 11, 403. 2019. [Online]. Available from: https://www.mdpi.com/2072-4292/11/4/403 [retrieved: 02, 2023].

[16] M. Wurm, H. Taubenböck, J. Goebel, T. Esch and, G. Wagner, "Census of the future? - Earth observation for spatial estimation of population distribution", in Standort - Journal for Applied Geography 35, pp. 169–175, 2011.

[17] H. Fan, A. Zipf, Q. Fu and, P. Neis, "Quality assessment for building footprints data on OpenStreetMap", International Journal of Geographical Information Science 28. S. 700–719. 2014.

[18] A. Herrmann, The reduction of land use through urban planning law. Legal validity and realization in the social system. Nomos Verlagsgesellschaft mbH & Co. KG. Baden-Baden, 2019.

[19] M. Kment, Public building law. Volume I Building planning law - Series of legal education, 107(7), C.H. Beck, Munich, 2017.

[20] M.-I. Ruprecht, "Small-scale settlement analyses for mapping changes: The utility of building-spatial data for planning", in Spatial Research and Planning, 72, pp. 227-238, 2014.

[21] Statistical Offices of the Federal and State Governments. Results of the 2011 Census for download - extended. [Online] Available from: https://www.zensus2011.de/DE/Home/Aktuelles/DemografischeGrunddaten.html [retrieved: 02, 2023].

[22] M. Neutze, "Grid-based evaluations of the 2011 census", in Stadtforschung und Statistik, pp. 64-67, 2015.

[23] E. Pajares, R. Muñoz Nieto, L. Meng and, G. Wulfhorst, "Population Disaggregation on the Building Level Based on Outdated Census Data". ISPRS International Journal of Geo-Information, 10(10):662, 2021. [Online] Available from: https://doi.org/10.3390/ijgi10100662 [retrieved: 02, 2023].

[24] D. Visca, M. Hoppe and, P. Neis, "Generating a Pseudo Resident Registration Register by Using Open Data", in GEOProcessing 2022: The Fourteenth International Conference on Advanced Geographic Information Systems, Applications, and Services. Porto, 2022.

[25] R.M. Olbricht, "Data Retrieval for Small Spatial Regions in OpenStreetMap", in J. Jokar Arsanjani, A. Zipf, P. Mooney, M. Helbich, Eds. OpenStreetMap in GIScience. Lecture Notes in Geoinformation and Cartography, pp. 101-122. 2015.

[26] H. Schröteler-von Brandt, "History of urban planning", in ARL (eds.): Handbook of urban and spatial development, pp. 805-821. 2018.

[27] G. Meinel, "The challenge of land monitoring: data sources for a land information system and what they can achieve", Stadtforschung und Statistik: Zeitschrift des Verbandes Deutscher Städtestatistiker, 33(1), pp. 107-114, 2020.

[28] S. Fina, "Urban Green Spaces Under Pressure: Comparative Values for Urban Green Space Provision in German Cities", Stadtforschung und Statistik: Zeitschrift des Verbandes Deutscher Städtestatistiker, 34(2), pp. 17-23. 2021.

[29] R. Hecht, C. Kunze and, S. Hahmann, "Measuring Completeness of Building Footprints in OpenStreetMap over Space and Time", ISPRS Int. J. Geo-Inf., 2, pp. 1066-1091. 2013.

[30] P. Neis and D. Zielstra, "Recent Developments and Future Trends in Volunteered Geographic Information Research: The Case of OpenStreetMap", Future Internet, 6, pp. 76-106. 2014. [Online]. Available from: https://www.mdpi.com/1999-5903/6/1/76 [retrieved: 02, 2023].

[31] COMMISSION REGULATION (EU) No 1089/2010 of 23 November 2010 Implementing Directive 2007/2/EC of the European Parliament and of the Council as Regards Interoperability of Spatial Data Sets and Services. [Online]. Available from: https://eur-lex.europa.eu/eli/reg/2010/1089 [retrieved: 02, 2023.

Towards Accurate Traceability of Water Reaching the Reservoirs

Ortega, L.
Department of Computer Science
University of Jaen
Jaen, Spain.
lidia@ujaen.es

Ramos, M.I.
Department of Cartographic Engineering, Geodesy and Photogrammetry.
University of Jaen
Jaen, Spain.
miramos@ujaen.es

Calle, A.
Researcher hired in project PYC20-RE-005-UJA
University of Jaen.
Jaen, Spain.
acm00225@red.ujaen.es

Abstract— **Drinking water is one of the natural resources most consumed by human beings and an important transmitter of diseases if it is not subjected to adequate quality controls. In Spain, much of the drinking water comes from reservoirs whose watershed is often occupied by farms. This means that part of the reservoir water has been in contact with phytosanitary products used on these farms. In this work, a methodology is being developed to determine the origin of the water flow that reaches certain areas of the reservoir. It is a 3D modelling and simulation methodology for the advance of the flow after a given rainfall event. To this end, work is being carried out on an improved version of the D8 algorithm, in which an extensive hydrological analysis is carried out, allowing more input parameters to be controlled in a more detailed manner. This allows 3D modelling and simulation of the rainwater trajectory until it reaches the reservoir.**

Keywords- Extended D8 algorithm; Drainage Network,; Chemical Monitoring.

I. Introduction

The use of crop protection products in agriculture is a common practice in agriculture to increase crop yields. However, the extensive use of these products mainly affects the soil, and sooner or later these substances end up in rivers and aquifers. In some countries in the Mediterranean basin, many rivers and aquifers flow into reservoirs whose water is also used for irrigation and human consumption. Soil conservation is therefore a key point for obtaining good agricultural yields and at the same time eliminating dependence on chemicals. Technicians and public administrations are looking for effective mechanisms to control these discharges and to carry out water traceability, especially in the case of water intended for human consumption. At present, the use of chemicals is not banned in agriculture or in the European Union (EU). They are widely used to fertilize the soil or to eradicate pests. In fact, it is difficult to find a balance between increasing production and preserving the environment. Food supply and food security are key objectives at EU and United Nations level.

A number of organizations have committed themselves in the coming years to address efficient soil management to meet the major challenge of increasing food production while minimizing soil degradation through sustainable development plans. In this regard, the 2030 Agenda and the Sustainable Development Goals (SDGs) adopted by all United Nations (UN) member states in 2015 call on all nations to combine economic prosperity, social inclusion and environmental sustainability with peaceful societies. They are closely linked to the Paris Agreement (United Nations, 2015) on climate change, which is incorporated into SDG 13 (Climate Action). These goals and targets entered into force on 1 January 2016 and will guide government decisions in the coming years.

The present work focuses on developing a methodology for the traceability of rainwater reaching a reservoir and determining its possible origin considering the crop areas of the basin it supplies. Specifically, the study focuses on the water reaching the Rumblar reservoir in the province of Jaen (southern Spain). This reservoir is mainly fed by river water from the Guadavelín and Pinto rivers. This reservoir supplies drinking water to a large part of the inhabitants of the province of Jaen. The water from this reservoir is subject to quality controls by governmental companies, which carry out regular chemical analyses. The company's technicians have observed that, after a period of rain, the chemical analyses reveal a higher concentration of nitrates. These components come mostly from olive groves, which are predominant throughout the catchment area. Therefore, an exhaustive study on the origin of the discharges should involve the study of the catchment basins, land use and orography of the terrain. Drainage networks have been used to study chemical tracing [1], as well as GIS techniques [2]. However, 3D simulations have also been used for similar purposes [3], but employed as stand-alone techniques.

In this work, an improvement of the D8 flow model algorithm is being developed in order to establish a workflow in which different scenarios can be simulated in the course of rainwater from its precipitation to its arrival at the reservoir. This simulation is also used to determine the concentration of substances at specific locations and to make a comparison with the agricultural use of the land. In this aspect, GIS tools provide the capacity to cross-reference all this information on the drainage network with the different land uses and to be able to focus the analysis on specific farms.

The document is structured as follows. Section 2 describes the methodology carried out and the data processing carried out jointly in QGIS and the D8 algorithm enhancement. Section 3 shows the first results with a plugin that has been implemented in Qgis to make it easier to use the extended D8 algorithm. Section 4 describes the conclusions and future work to be developed.

II. METHODOLOGY

This section describes the process to obtain the traceability of water runoff along its path to the reservoir. The objective is to know which areas the rainwater comes from and arrives at specific points in the river basin that finally discharge their water into the reservoir. The aim is also to obtain the percentage of water that each of these areas contributes to the points under study. It is important to know whether this origin is agricultural or not in order to be able to carry out a more detailed study of the areas that contribute water to the reservoir. Therefore, this study helps to identify which farms contribute to the concentration of discharges or make inappropriate use of fertilizers or plant protection products. Fig. 1 shows schematically the process. The first two steps identify a series of representative points on which to focus the study and in the second phase (steps 3 and 4) a detailed study of the traceability of the water flow in these representative points will be carried out.

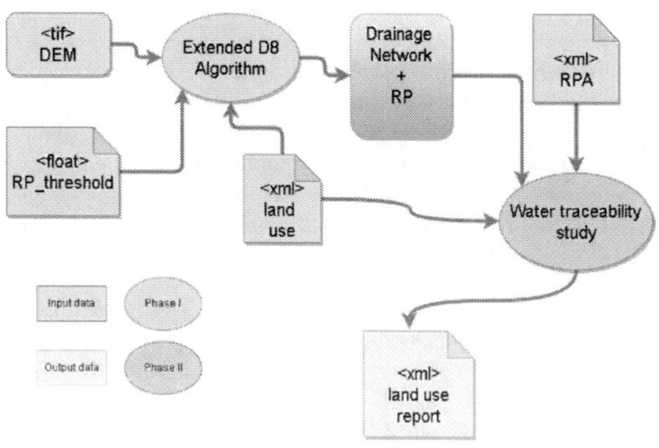

Figure 1. Flow diagram of the methodology proposed.

A. Research Area

The study was carried out in the area around the Rumblar reservoir, which is a storage reservoir built on the course of the river of the same name (38° 09′ 42″ N 3° 48′ 15″ W, ETRS89), in the municipality of Baños de la Encina, in the province of Jaen, southern Spain. This region has a semi-continental climate with a mean annual temperature of 25.1°C, a mean relative humidity of 70% and a mean annual precipitation of 1341 mm. The capacity of the reservoir is 126 hm^3 although it is currently at 16 % of its capacity [4]. The reservoir is located in the heart of the Sierra Morena mountain range, surrounded by Mediterranean woodland with holm oaks and cork oaks, and is an ideal habitat for species such as the Iberian lynx. In the surrounding area it is possible to observe a large group of birds of prey and carrion-eating birds, as well as some mammals such as otters, foxes and huntable species such as mouflon, fallow deer, wild boar and deer. A few kilometers from the reservoir, the relief and vegetation change considerably, with olive groves alternating with hilly pastures where wild cattle graze. For the purpose of our research, we covered an area of 56497 ha. Fig. 2 shows the location where the research has been carried out.

B. Data set

In the first phase of the work, the hydrographic elements of the study area were obtained from the Digital Elevation Model with a spatial resolution of 2x2m that forms part of the Spanish Spatial Data Infrastructure (IDEE) platform which follows INSPIRE Regulation (EC) No 1205/2008. This model was downloaded from the Download Center of the Autonomous Section of the National Center for Geographic Information [5]. It is a model obtained by interpolation from the terrain class of the LIDAR flights of the second coverage of the National Aerial Orthophotography Plan (PNOA) [6]. Another important layer of information to be integrated into the system and which will allow the identification of areas to monitor the dumping of phytosanitary products is that of land use in the Rumblar basin. The land uses established in the information layer have been obtained from the information compiled by CORINE Land Cover, a cartographic inventory of land use and a basic instrument in environmental and land-use planning policy in the EU.

C. D8 extended

The D8 method for obtaining the drainage network needs the DEM (Digital Elevation Model in tiff format) file of the area associated with the reservoir basin. This process works similarly to the D8 algorithm introduced by O'Callaghan and Mark [7]. Initially, the method codifies in each cell C(i,j) one of the eight directions where C(i,j) drains its water accumulation. Then, an iterative process makes in parallel that all cells drain their accumulated water to its neighboring cell, depending on this specific direction. The process finishes when no more water is drained along an entire iteration. Those specific points whose accumulated water exceeds a threshold (RP_threshold in Figure 1) at any moment of the process, are considered representative points (RP). The set of RPs is the input of this first process, which in fact works similarly to the original D8 method. However in a second phase, the algorithm is processed again and for each representative point and for each iteration of the process, the following information is stored in a set of tuples <id-plot is the identification number of the plot and accumulated-water the amount of water from this plot>.

There is one tuple for each different plot pouring water sooner or later into this RP. The plot is represented by a specific identification number, and for each of these plots we maintain information about the water traversing the RP. The algorithm finds these representative points (RP) automatically, however the user can add specific ones (RPA) if additional strategic locations are decided to be used for chemical analysis. This information is the input for further processing that can be done in QGIS.

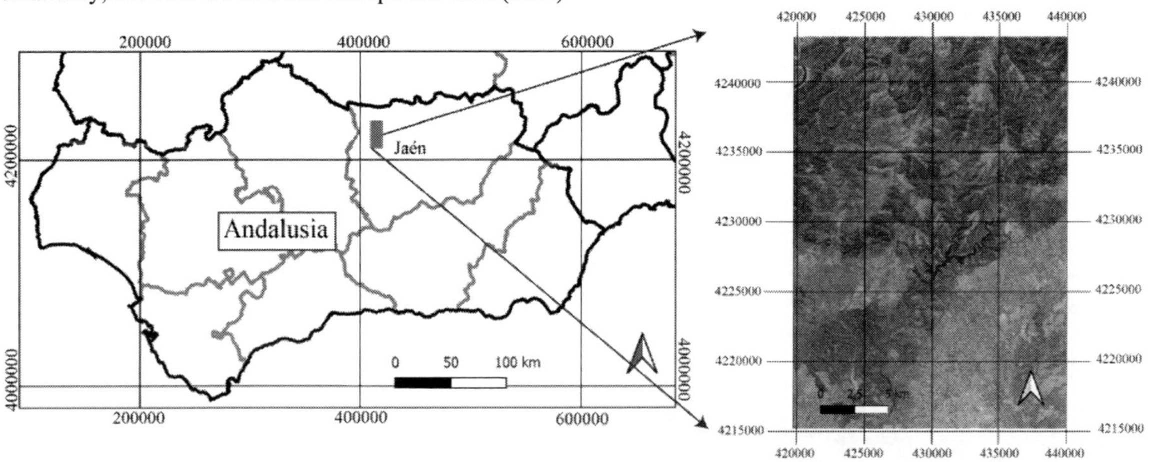

Figure 2. Location of the research area. The left image represents the south of Spain (Andalusia). The right image is the map of the Rumblar reservoir in Jaen. The coordinates (m) are UTM, zone 30 referred to ETRS89.

III. RESULTS

Once the extended algorithm has been implemented, the idea is to be able to identify a drainage network according to different previously established conditions, Fig.3.

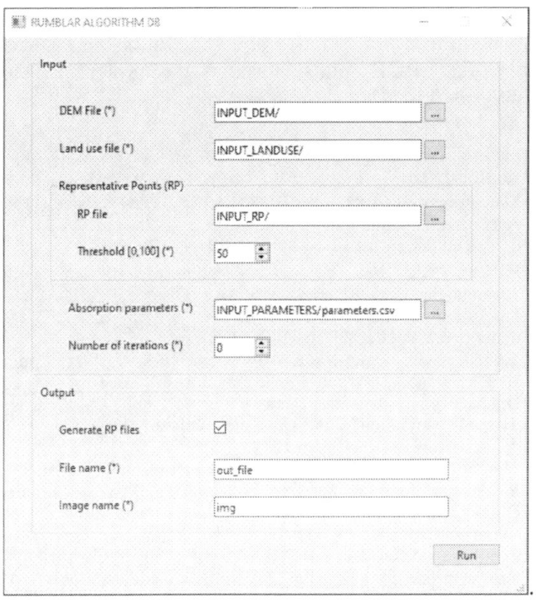

Figure 3. Interface to the main menu of the system in which the extended D8 algorithm has been implemented.

First of all, our algorithm needs the DEM and the land use layer of the working area, which is necessary information to be able to carry out the traceability analysis, which is the main subject of this work. Secondly, an interesting functionality is to be able to add new RPs to the system, not only those generated by the classical D8.

On the other hand, the classical D8 algorithm simulates the same amount of precipitation at each pixel. However, in the extended version of the algorithm, the amount of precipitation is also an input parameter depending on the areas. From the zonal cumulative precipitation maps, each pixel can be assigned a different precipitation value, which is very interesting when working with large areas of land, as the amount of precipitation is different in each zone. It is also possible to define different levels of water absorption by the soil, depending on the underlying soil type. The absorption capacity of the soil depends not only on the slope extracted from the DEM, but also on the type of material on which it rains.

The number of iterations is another input variable of the extended algorithm. This parameter controls the whole execution process of the algorithm. By default it runs completely, until the rainwater runoff ceases. But it is possible to set a certain number and control the progress of the process at intermediate stages. It will also be possible to overlay land use and query the system to see which areas have received the most rainwater. An Qgis plugin has also been implemented to allow interaction between the user and the system. The system automatically selects on the map the plots that have contributed water to that point, and the percentage that it represents with respect to the total water accumulated in it.

In this case, a provisional result can be displayed. Fig. 4, which shows a screenshot in which it can be seen how, when clicking on a characteristic point, the algorithm identifies

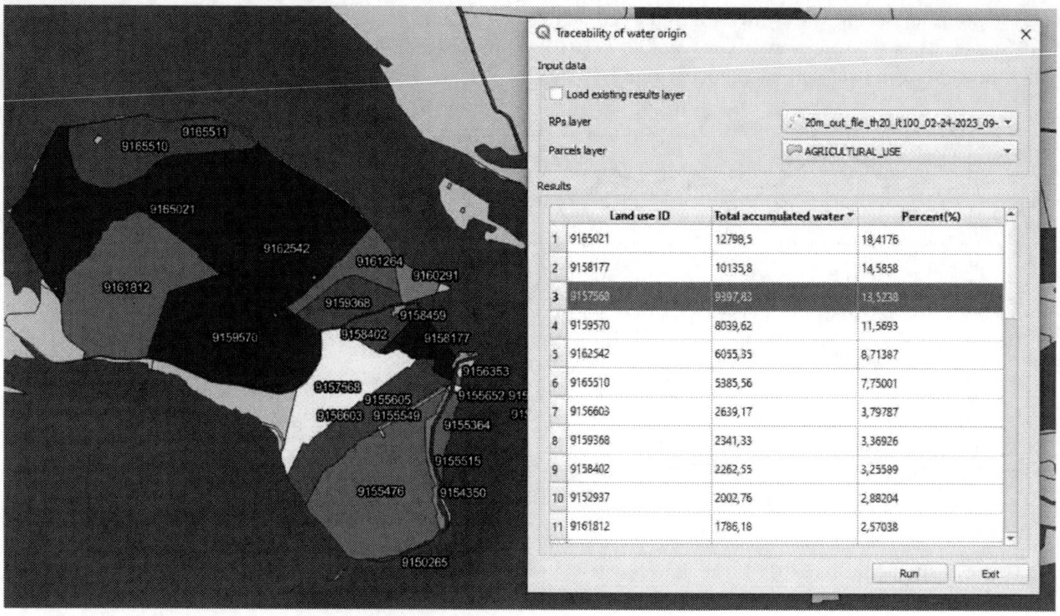

Figure 4. Example of a query executed with the Plugin.

from which plots the water that has reached that point comes from, and the information related to each plot. This information is essential for water treatment plant managers, since by analysing water samples in specific areas, if they detect abnormal values of certain chemical substances, they will be able to know from which farms these pollutants may be arriving.

Finally, it will be possible to consult from which areas rainwater reaches the reservoir and which are absorbed by the soil and do not reach it. It is therefore a question of having a system in which we can use different configuration parameters of the drainage network to achieve the most accurate traceability possible of the water that reaches the reservoir, with the added value of using an easy-to-use application.

IV. Conclusions

In this work, a methodology has been developed to determine the origin of the water flow that arrives at certain points in the reservoir basin. It is a methodology for modelling and simulating the advance of the flow once it has rained using an improved version of the D8 algorithm. The resulting developed work is an easy to use tool by managers of drinking water treatment stations. The spatial information uploaded into a GIS software joined to the run of the extended D8 algorithm helps us to extract information about the study area with just one click on the screen. The algorithm is still undergoing debugging and validation. It needs to be tested against field data in different parts of the study area. However, the tests carried out so far show promising results.

Acknowledgment

This research has been partially funded through the research projects PYC20-RE-005-UJA, 1381202-GEU and which are co-financed with the European Union FEDER and the Junta de Andalucía funds. Also partially supported by the Spanish Ministry of Science and Innovation, the European Union (via ERDF funds) through the project PID2021-126339OB-I00.

References

[1] Williams, M.R.; King, K.W.; Fausey, N.R. "Drainage Water Management Effects on Tile Discharge and Water Quality". Agricultural Water Management, vol. 148, pp. 43–51, 2015, doi:https://doi.org/10.1016/j.agwat.2014.09.017.

[2] Habeeb, N.J.; Weli, S.T. "Combination of GIS with Different Technologies for Water Quality: An Overview". HighTech and Innovation Journal vol. 2, n. 3, pp. 262–272, 2021, doi.org/10.28991/HIJ-2021-02-03-1.0.

[3] Karandish, F.; Darzi-Naftchali, A.; Šimůnek, J. "Application of HYDRUS (2D/3D) for Predicting the Influence of Subsurface Drainage on Soil Water Dynamics in a Rainfed-Canola Cropping System". Irrigation and Drainage, vol. 67, pp. 29–39, 2018 doi:https://doi.org/10.1002/ird.2194.

[4] Ministerio de agricultura, pesca y alimentación Digitisation Strategy for the Agri-Food and Forestry Sector and Rural Areas; Estrategia de Digitalización del Sector Agroalimentario y Forestal y del Medio Rural; 2020.

[5] Centro de Descargas del CNIG (IGN) [Online]. Avalilable from http://centrodedescargas.cnig.es/CentroDescargas/linkUnMD 2022.07.06.

[6] Plan Nacional de Ortofotografía Aérea [Online]. Available from http://pnoa.ign.es/estado-del proyecto-lidar. 2022.07.06.

[7] O'Callaghan, J.F.; Mark, D.M. "The Extraction of Drainage Networks from Digital Elevation Data". Computer vision, graphics, and image processing vol. 28, pp. 323–344. 1984. doi.org/10.1016/S0734-189X(84)80011-0.

Terrestrial Laser Scanner High Station to Control the Quality of DEM Data

Juan F. Reinoso-Gordo
Dept. Expresión Gráfica Arquitectónica y en la Ingeniería
University of Granada, Spain
Email: jreinoso@ugr.es

Francisco J. Ariza-López, Antonio Mozas-Calvache, José L. García-Balboa, J. Ruiz-Lendínez
Dept. Ingeniería Cartográfica, Geodésica y Fotogrametría
University of Jaén, Spain
Email: fjariza@ujaen.es, antmozas@ujaen.es, jlbalboa@ujaen.es, lendinez@ujaen.es

Abstract— **Currently many Digital Elevation Models (DEMs) are derived from surveys obtained by LiDAR flights. The quality of both products is usually assessed using control point-based techniques. Although it seems that this way of reporting quality is not fully adequate because the superficial nature of both the DEMs and the area covered by the LiDAR survey. For this last reason (the superficial nature of the objects to be controlled), according to the ideas contribution in Geoprocessing we propose to control the quality of the DEMs and LiDAR flights by means of a Terrestrial Laser Scanner (TLS) located on a pole at 6 or 7 m from the ground. We propose a configuration of 4 scan stations registered in a single point cloud and georeferenced, so that its accuracy is greater than the product that is intended to be controlled.**

Keywords- LiDAR; DEMs; quality; terrestrial scanner laser; georeference; accuracy

I. INTRODUCTION

It is very important to know the Light Detection And Ranging (LiDAR) surveys quality and the DEMs derived from them. These two types of geospatial data are the basis to develop a large number of products in sciences, such as Civil Engineering, the Environment, Hydrology, Geology, etc. But in order to derive quality products, such as drainage networks, flood zones, slope maps, etc. it is necessary to know the quality of the original data (LiDAR and DEMs). And to report the quality of LiDAR and DEMs it is necessary to carry out expensive quality controls [1]. Traditionally, quality assessment has been based on control points, as for example shown in [2], where conventional parameters are used (Root Mean Square Error RMSE), μ, σ, Normalized Median Absolute Deviation (NMAD), etc.), [4]. Many of these parameters assume normality in the data distribution, which is not usually true. This point-centered approach allows for numerous positional control methods [5], one of the most recent being the one proposed in [2], although on many occasions it requires the point to be controlled to be well identified. Additionally, if it is intended to evaluate the quality of a DEM, which has a superficial nature, it seems more appropriate that its positional accuracy be evaluated and controlled by a sample whose elements are also superficial. For this reason, in this study we propose to use patches to study the quality of both LiDAR and DEMs data. This approach is novel, since we have not found anything similar in the reviews we have carried out [6] on the study of the quality of both LiDAR or DEMs data. To test the application of this methodology, we propose to use an existing LiDAR flight that is controlled by a survey with TLS stationed on a pole at 6 and 7 m from ground. This greater height of the TLS setup tries to get a greater perpendicularity between the incident rays and the ground and consequently a lower precision dilution.

The advantage from the above proposal compared to the traditional control points approach is that the patch includes more information than only one point, and the corresponding point cloud is able to derive different statistics that can be used in simulation parametric models.

In section II it is explained the method and material used in this research from the LiDAR product to the reference captured data. In section III the results are shown and some comments about the method proposed and linked to accuracy are included.

II. MATERIAL AND METHOD

As a product to be controlled, an existing LiDAR flight will be used, which we will call (Mpro). This model was an experimental LiDAR flight with a density of 14 points per square meter as shown in Figure 1.

Figure 1. LiDAR sample for the Mpro

The Reference Model (Mref) Figure 2 will be obtained by means of a Leica TLS (BLK360) whose distance measurement accuracy is 6 mm according to its specifications. The BLK360 will be mounted on a pole 6 or 7 meters high and stabilized by a tripod that can be extended up to 4.5m. Usually it will not be necessary to use the tripod maximum extension; only in windy situations that can cause the pole to flex will the tripod extend to its maximum length.

Figure 2. Mref from the TLS stations registration

The scanner was placed on the pole in an inverted position so that no shadow areas were created due to the scanner base, to get that setting an arm adapted to the pole was used to hold the scanner Figure 3.

Figure 3. BLK36 scanner mounted on pole in tripod

It is intended to take an area with such an extension that it allows to extract a square surface of 50x50 m; those dimensions will be the patch size that will be used to control both the LiDAR point cloud and its derived DEM. To get the patch, 4 scan stations are used divided in two groups. A group is composed of 2 scans where the vertical axis of the pole remains fixed while the position of the second scan in the group is achieved by turning the pole 180° on its own axis. In Figure 4 2 yellow pyramids are shown at one end of the line: the pyramids represent the 2 scan stations corresponding to group 1. At the other end of the line, the 2 scans of group 2 will be located.

Figure 4. Scan stations from the same group (yellow pyramids)

The 4 scans were georeferenced and registered using the Cyclone Register 360 software, which required the coordinates of several targets, obtained through GNSS RTK in differential mode connected to the server through Ntrip. The Figure 5 shows the targets distribution and the errors obtained after registering.

Figure 5. Registration using target coordinates

III. RESULTS AND CONCLUSIONS

After filtering the registered point cloud we get the Mref that could be compared to the Mpro in order to compute the Mpro quality. As shown in Figure 6, the set error in every registration is smaller than 0.03 m and the RSME for target points is smaller than 0.04 m (Mref). Because the RSME for the Mpro is 0.15 m, according to the LiDAR flight specifications, the registered and filtered Mref is suitable for controlling the Mpro LiDAR quality.

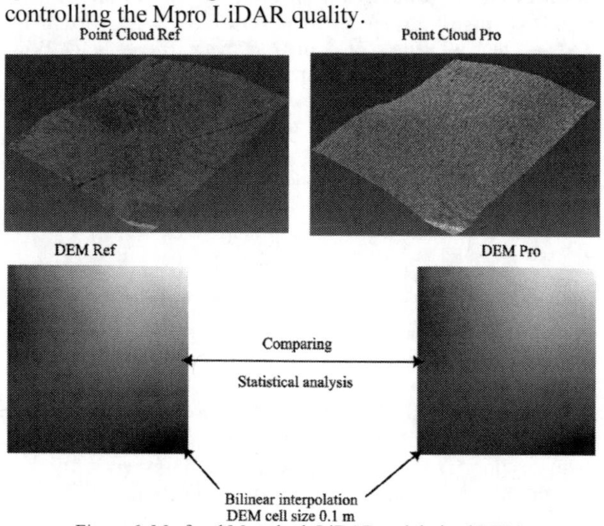

Figure 6. Mref and Mpro both LiDAR and derived DEMs

Our proposal, also include controlling the DEM quality starting from the Mpro and Mref point cloud. Before computing the DEMpro quality it is necessary determine a method to pass from Mpro and Mref to DEMpro and DEMref. We propose a bilinear interpolation to derive both DEMs with a cell size of 0.1 m.

The quality control can be performed from two point of view (altimetric and planimetric). The altimetric quality can be analyzed through the cumulative error distribution function and the planimetric one can be addressed by contours as proposed in [7].

The data used in this study was the experimental LiDAR flight from the area of Navarra in Spain (10391 km^2) and the futures plans are to extend the methodology to Andalusia in Spain (87599 km^2).

REFERENCES

[1] Y. Bühler, M. Marty, C. Ginzler, High resolution DEM generation in high-alpine terrain using airborne Remote Sensing techniques. Trans. in GID, vol. 16(5), pp. 635-647, 2012.

[2] ASPRS, ASPRS Positional Accuracy Standards for Digital Geospatial Data, PE&RS, 2015.

[3] P. Zandbergen, Positional Accuracy of Spatial Data: Non-Normal Distributions and a Critique of the National Standard for Spatial Data Accuracy. Transactions in GIS, 12(1), pp. 103–130, 2008.

[4] D. Maune, Digital elevation model technologies and applications. The DEM user's manual. ASPRS, 2007

[5] F. J. Ariza and A. D. Atkinson, Analysis of some Positional Accuracy Assessment Methodologies. Surveying Engineering, 134(2), pp. 45-54, 2008.

[6] J. L. Mesa-Mingorance and F. J. Ariza-López, Accuracy Assessment of Digital Elevation Models (DEMs): A Critical Review of Practices of the Past Three Decades. Remote Sensing, 12(16), pp. 1-27, 2020.

[7] J. F. Reinoso, An algorithm for automatically computing the horizontal shift between homologous contours from DTMs. ISPRS Journal of Photogrammetry and Remote Sensing, vol. 66 (3), pp. 272-286, 2011

Using a Social Network for Road Accidents Detection, Geolocation and Notification - A Machine Learning Approach

Samuel Pereira de Vasconcelos
Information Systems Laboratory
Federal University of Campina Grande
Campina Grande - PB, Brazil
email: samuel.vasconcelos@ccc.ufcg.edu.br

Cláudio de Souza Baptista
Information Systems Laboratory
Federal University of Campina Grande
Campina Grande - PB, Brazil
email: baptista@computacao.ufcg.edu.br

Hugo Feitosa de Figueiredo
Information Systems Laboratory
Federal Institute of Technology of Paraiba
Esperança – PB, Brazil
email: hugo.figueiredo@ifpb.edu.br

Abstract - **The road system is the main means of transport used in Brazil. Traffic accidents are quite common in this mode of transport, incurring one of the biggest causes of death in the country. Profiles on social networks of the Federal Highway Police (PRF) and other sources of information, contribute to alert drivers as quickly as possible about road accidents that have occurred, in order to prevent other accidents from occurring. Also, such information can be used by drivers about possible delays, or even deviations in their paths. However, accessing such information via text while driving is illegal and further increases the risk of accidents. Therefore, this paper addresses a study about reliable posts in social networks, in particular Twitter, to create a supervised classification model, which is capable of classifying tweets about the occurrence or not of accidents. The results include the best induction model obtained for classifying tweets, among several analyzed, as well as the construction of a mobile application that can notify through audio drivers about accidents reported on their way, in real time.**

Keywords - Road accidents; Machine learning, Natural Language Processing; Geoprocessing; Mobile Computing; Social media.

I. INTRODUCTION

Road accidents have been one of the most important challenges of contemporary society worldwide, being the cause of concern and studies in several countries. Road accidents are the eighth leading cause of death in the world [1].

On the other hand, we live in a highly connected world. Social networks may help in disseminating traffic accident alerts in order to attract the attention of those who are traveling near the accident hotspot in order to prevent further accidents.

Thus, developing technologies that allow road accidents alerts to reach, in real time and with reliability, the drivers who will cross such an accident is a challenge of socio-economic and public health interest, which gives rise to the focus of this work.

There are several channels on the Internet to alert the population about accidents, most of them in the form of text (e.g., tweets), which makes it impossible for drivers while traveling to access them. Also, there are applications for smartphones such as waze that aims to alert drivers while traveling on the highways; however, such applications are not completely reliable, as it comes from crowdsourcing [2]. Hence, using official communication channels, such as the Twitter profiles from the government, avoids misinformation.

This work proposes a framework that can automatically identify, monitor and alert drivers through audio and in real time about the incidence of road accidents using machine learning, geoprocessing and Natural Language Processing techniques based on tweets from authorities, such as the Brazilian Federal Highway Police (PRF).

The remainder of this paper is structured as follows. Section II discusses some related works on road accidents. Section III presents the methodology used in this research. In Section IV, the experiments carried out to identify the best machine learning model for the purpose of this research are discussed, as well as the analysis of the results obtained. Finally, Section V concludes the paper and presents future guidelines to continue this research.

II. RELATED WORK

There are several works that address different aspects of road accidents around the world. There are studies on the social, economic and environmental impacts of road accidents [3][4]. Other studies are focused on victims of road accidents, whether fatal or even sequelae [5][6]. Some authors are concerned with detecting potential patterns in road accidents [7]-[11]. Also, there are studies aimed at analyzing the main causes of road accidents [12]-[18]. Finally, there are studies aimed at classifying and predicting the severity of accidents [19]-[27].

Katsoukis et al. [28] used data mining techniques to classify the risk of accidents in regions in Greece, considering the number of accidents in a given region. Ryder and Wortmann [29] proposed an approach to detect and classify places with a high rate of accidents; aiming to alert drivers in real time about imminent dangers on the highways, through a mobile application.

Ryder et al. [29] developed a decision support system to prevent road accidents. This system sends alerts to drivers when they are close to high accident risk areas.

Ren et al. [30] proposed a deep learning model, through a neural network Long Short-Term Memory (LSTM), based on spatiotemporal data correlation in order to predict traffic accident areas in Beijing.

Zhang et al. [31] proposed a method to identify key points where the incidence of road accidents is high in a given period of time. The dataset contains characteristics such as: holiday, day of the week, time, crash site type, and weather conditions. Yu et al. [32] proposed a neural network model to predict accidents on highways, called STEEN, which combines spatial distributions, temporal dynamics and external factors, such as Points of Interest (POIs).

To the best of our knowledge, there is no work that simultaneously uses machine learning techniques, natural language processing and geoprocessing for the problem of detecting road accidents and their communication to drivers. The BERT transformer obtained the best results in our experiments. We experimented several machine learning models based on supervised classification. We use NLP regular expressions and preprocessing techniques. Finally, we georeferenced tweets from texts and store them in a spatial database system.

This integrated solution consists of the main contribution of this paper.

III. METHODOLOGY

The methodology used in this research was based on the Cross Industry Standard Process for Data Mining (CRISP-DM) [33]. This methodology consists of six steps: (i) Business Understanding, (ii) Data Understanding, (iii) Data Preparation, (iv) Modeling, (v) Evaluation and (vi) Deployment.

The Business Understanding step was covered in Section I. The Data Understanding step consisted of identifying the PRF profiles on Twitter and analyzing the keywords to be used to gather the tweets. The Data Preparation step addressed pre-processing and data labeling to create an annotated corpus. The Model Induction step focused on deploying several induction models, using supervised classification, with hyperparameter tuning. The Evaluation step carried out the analysis of the results of the models that were generated in the previous stage, choosing the model that presented the best performance. Finally, the Deployment step consisted of implementing the model in the SafeTrip tool in a mobile platform.

In the following, we detail the methodology steps.

A. Data Understanding

This step consists of selecting the Twitter PRF profiles; the choice of keywords used in road accidents; as well as the tweet gathering process.

1) Twitter Profile Selection: During the search for Twitter profiles, to avoid data poisoning, official PRF profiles from several states in Brazil were chosen, due to reliability on reporting road accidents. The chosen profiles were all active by November 2021 including: @PRFParana, @PRFCeara, @PRF191RJ, @PRF191PR @PRF191TOCANTINS, @PRF191SP, @PRF191SERGIPE, @PRF191RORAIMA, @PRF191RONDONIA, @PRF191PE, @PRF191PA, @PRF191MS, @PRF191ES, @PRF191AM, @prf_sc, @PRF191ACRE, @prf_rn, @prf_pi, @prf_pb, @PRF_MS, @prf_mg, @prf_df, @prf_ba e @prf_al.

2) Keyword Selection: We chose keywords commonly used in posts about road accidents in Twitter PRF profiles. The words chosen were: 'wounded', 'accident', 'death', 'shock', 'tragedy', 'run over', 'overturn', 'victim', 'collision', 'turn', 'turned', ' pileup', 'fire', 'crash', 'tip over', 'crash', 'left the runway', 'crashed', 'dies', 'passed away' and 'fell over'. Figure 1 depicts an example of tweet from the Brazilian Federal Highway Police (PRF) of the State of Paraiba.

Figure 1. Example of a tweet with a highway accident alert.

3) Tweets Gathering Process: After choosing the profiles and keywords related to traffic accidents, requests were made to the Twitter API. The data returned by the Twitter API include: identification number, user, date, place of publication, text, and likes.

B. Data Preparation

This step carried out the manual data labeling and tweet pre-processing.

1) Data Labeling: We performed the manual labeling of tweets as follows. We label a positive class when the content contains information about traffic accidents, including the location where the incident occurred; whereas we label the negative class when the tweet did not contain data on traffic accidents. The resulting corpus has a structure similar to Table I.

TABLE I
TWEET CLASSIFICATION EXAMPLE.

Tweet	Label
"the Dom Pedro road has stretches with rain this afternoon, drivers should redouble their attention, works that were carried out this friday have already been closed, there are no accidents and traffic flows well"	NEGATIVE
"attention br 116 km 643 jequie road partially closed due to accident reduce speed"	POSITIVE

At the end of the data labeling process, 3,311 tweets were obtained, but the dataset was quite unbalanced, the negative class occupied about 75% of the records, as shown in Figure 2.

2) Data Preprocessing: The collected data were preprocessed using several NLP techniques . First we converted text to lowercase, and removed non-relevant information such as links, emoticons and punctuation from the tweets using regular expressions. Also, we performed tokenization and removed stopwords.

We also performed corpus balancing to avoid possible bias in the classification models to be used. For that we used the statistical technique of subsampling [34]. Thus, the balanced set resulted in 832 instances for each class.

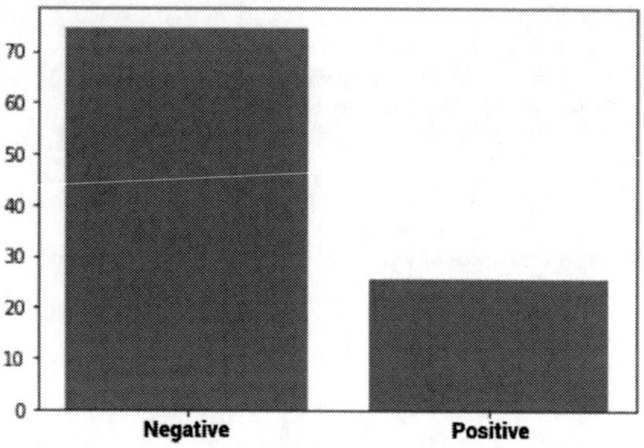

Figure 2. Labeled tweets distribution.

IV. Model Induction and Evaluation

In order to find out the best model that fits our solution, we ran several classification algorithms and compared their results.

A. Model Setup

To obtain the best supervised classification model for our proposed problem we used the following machine learning techniques: Naive Bayes, Logistic Regression, XGBoost, Support Vector Machine (SVM), Random Forest and the BERT transformer.

Our dataset was split into training and testing, 20% of the data were allocated to the test set, while 80% were allocated to training, with 10% of this training set being used during training as a validation set. To reduce the chances of overfitting, we used the k-folds cross-validation method with the parameter k = 10.

We used the sklearn GridSearchCV algorithm for tuning the model's hyperparameters o during training. We used accuracy, precision, recall and f1-score metrics to evaluate the models performance. Different combinations of hyperparameters were tested in order to find the model that best maximized the accuracy and f1-score metrics. The corpus and all the code used for training the models can be accessed through *google colaboratory* [35].

Concerning the transformer classifier, we used BERTimbau. BERTimbau is a pre-trained neural network to deal with Portuguese. For the model induction, pytorch tensors were used for 4 epochs, number recommended by the literature, using the same training set of the sklearn models. The code and corpus used can be accessed through *google colaboratory* [36].

B. Model Evaluation

In this subsection, we present and analyze the results of the classification models in order to choose the best induction model for the road accidents classification problem.

Table II presents the metrics from the several classification algorithms used. BertTimbau was the best induction model.

TABLE II
MODELS EVALUATION.

	Métrics			
Model	Accuracy	Precision	Recall	F1
RandomForestClassifier	91,89%	87,36%	97,55%	92,17%
LinearSVC	92,19%	88,89%	96,39%	92,49%
MultinomialNB	93,69%	89,20%	98,74%	93,73%
LogisticRegression	94,89%	92,77%	96,86%	94,77%
XGBoost	95,50%	93,75%	97,63%	95,65%
BERTimbau	**97,29%**	**95,88%**	**98,79%**	**97,31%**

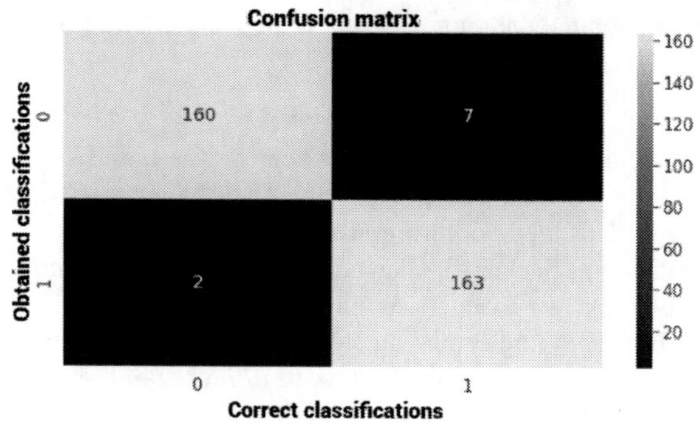

Figure 3. BERTimbau confusion matrix.

The confusion matrix for the BERTimbau model is presented in Figure 3.

To verify the absence of overfitting, Figure 4 plots the error behavior during training with training and validation data. Hence, we can observe that the validation data error decreases as the training one also decreases. Therefore, we can conclude that the model generalizes well.

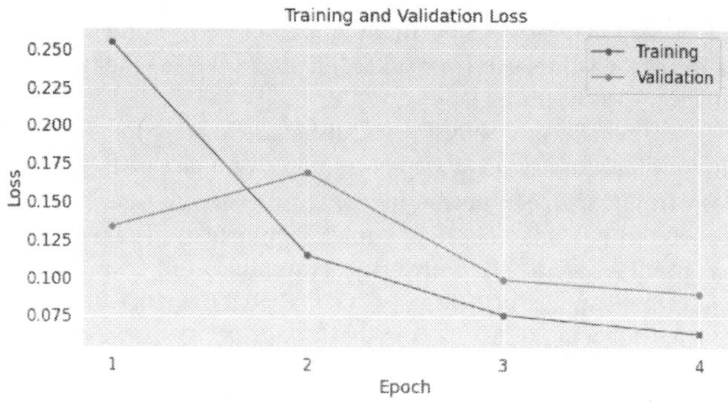

Figure 4. Overfitting test.

C. Deployment: Road Accidents System

The infrastructure of the accident alert system was developed with the help of the Python 3.10.2 programming language and the Postgresql 9.6 database for data management. In addition, the Spatial Postgis 3.2 extension was used.

The SafeTrip architecture is composed of three main components shown in Figure 5. The components are the crawler, the classifier and the mobile application.

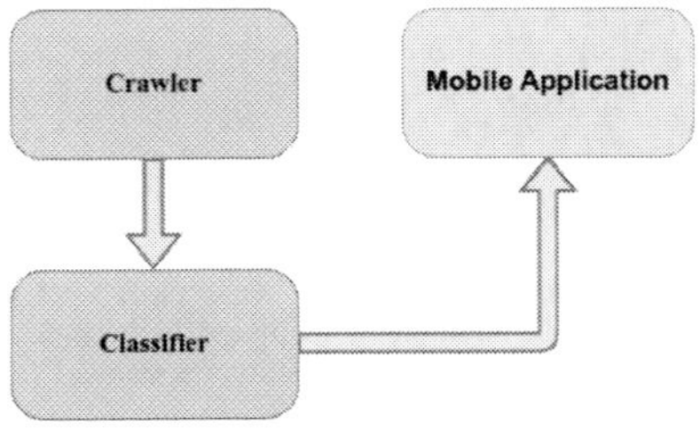

Figure 5. SafeTrip overall architecture.

1) Crawler: The crawler is responsible for tweets gathering.. With the help of the tweepy API, tweets were periodically collected, preprocessed and stored in the database, as shown in Figure 6.

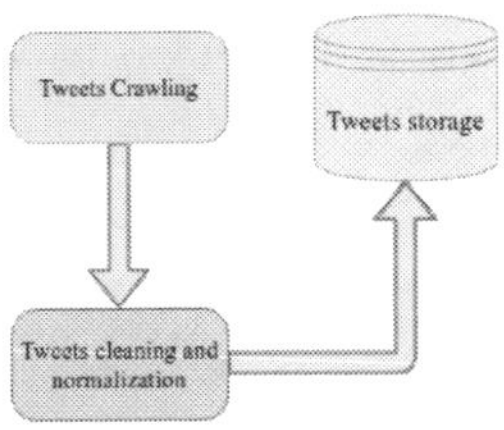

Figure 6. Crawler architecture.

2) Classifier: Figure 7 depicts the classifier module that is responsible for retrieving the tweet stored by the crawler and applying tokenization and data normalization to infer the tweet class according to the chosen classification model. When the tweets are classified as positive, a script based on regular expression is applied to extract information such as federal unit, code and kilometer of the highway to retrieve the accident geolocation. This geolocation consists of calculating the starting point of the kilometer where the accident occurred, so an algorithm was developed, described below and implemented in a stored procedure that, with the help of the database of Brazilian highways maintained by the Ministry of Infrastructure and the information extracted from the tweets, determines the latitude and longitude coordinates of the accident. At the end, this information, plus the time of occurrence and the text itself, are stored in the geographic database in the form of an alert.

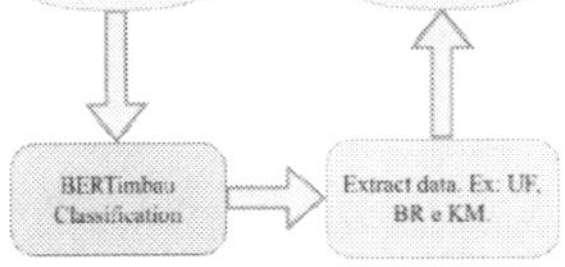

Figure 7. Classifier architecture.

3) The SafeTrip Mobile Application: In order to validate the results obtained and make them accessible to the community, a mobile application was developed to notify drivers by audio about accidents detected on Twitter. This mobile application was developed based on the client-server architecture model. According to Vaskevitch [37], in this architecture, processes are separated into independent platforms, allowing communication between processes while obtaining the maximum benefit from each different device. The server is responsible for monitoring PRF profiles, collecting published tweets, processing, classifying, extracting information, geolocating and sending alerts to users via websocket.

With the help of React Native, which is a javascript-based library that builds native code for Android and IOS applications, a mobile application was developed to act as a client, providing geolocation data and issuing alerts. This mobile application acts as a client in our architecture. Permissions are required to access the internet and also the location system of the device on which it is installed. In the mobile application home, users can choose one of all Brazilian cities as the destination city. The path between the user's current location and its destination will be displayed after the city is selected and the "Continue" button is pressed.

The client keeps the server informed about its location and when it detects an accident alert on the user's route, the server notifies the application that represents this information through an icon in the form of an exclamation mark, as shown in Figure 8. In addition, the textual content of the tweet that generated the alert is converted into audio.

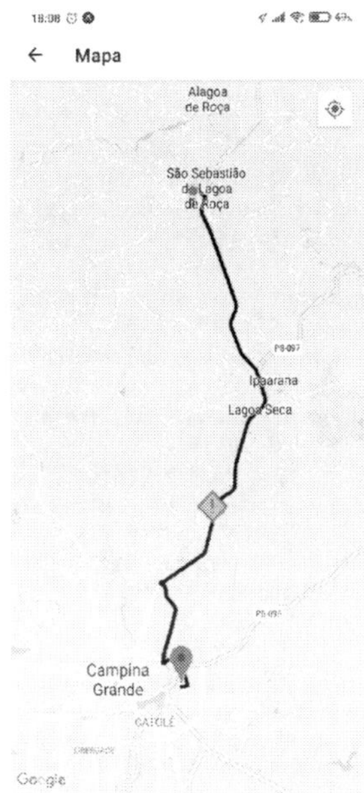

Figure 8. User´s route and accident allert.

V. CONCLUSION AND FUTURE WORK

This article addressed the issue of road accidents and how, using technologies based on machine learning, natural language processing, geoprocessing and mobile applications, we can mitigate such accidents and improve drivers' travel safety through sound alerts of accidents that occurred on the respective routes. For that, tweets from the Brazilian Federal Highway Police were used, which report, among other things, the occurrence of road accidents, with a very high reliability, which differs from the information reported in traffic monitoring applications that are based on crowdsourcing.

One of the main challenges in carrying out this research was the manual labeling of the dataset for training the supervised classification models. Another major challenge was trying to understand the function of each hyperparameter in the classification models and choosing those that presented the best results. Finally, in the area of geoprocessing, we had the challenge of transforming an accident reported in the form highway kilometer to latitude and longitude coordinates. The BERT transformer obtained the best results in our experiments.

As future work, we intend to increase the corpus of accidents, using data from other secure sources such as tweets from highway concessionaires and the press, as well as carrying out usability tests of the mobile application. In addition, with reliable tweet classification models, it is possible to contribute as one more source for real-time updating of the Brazilian Federal Highway Police database referring to the history of accidents, in its open data portal, since there is a delay of about 2 months in the data update.

ACKNOWLEDGMENT

The second author would like to thank the National Council for Scientific and Technological Development (CNPq), Brazil, for partially funding this research.

REFERENCES

[1] C. D. Mathers, T. Boerma, D. M. Fat, Global and regional causes of death, British Medical Bulletin, Volume 92, edition 1, pp. 7–32, 2009, https://doi.org/10.1093/bmb/ldp028.
[2] Google, Sobre o Waze, Available at https://support.google.com/waze/answer/6071177?hl=pt-BR&ref_topic=9022747, Accessed in 2023-03-09.
[3] A. Kocatepe, M. B. Ulak, E. E. Ozguven, and M. W. Horner,"Who might be affected by crashes? Identifying areas susceptible to crash injury risk and their major contributing factors", Transportmetrica A: transport science 15, pp. 1278–1305, 2019.
[4] C. Liu and A. Sharma, "Using the multivariate spatio-temporal Bayesian model to analyze traffic crashes by severity", Analytic methods in accident research 17, pp. 14–31, 2018.
[5] P. Tiwari, H. Dao, and G. N. Nguyen, "Performance evaluation of lazy, decision tree classifier and multilayer perceptron on traffic accident analysis", Informatica 41, 2017.
[6] P. Tiwari, S. Kumar, and D. Kalitin, "Road-user specific analysis of traffic accident using data mining techniques", In International Conference on Computational Intelligence, Communications, and Business Analytics, Springer, pp. 398–410, 2017.
[7] S. R. Santos, C. A. D. Junior, and R. Smarzaro, "Analyzing traffic accidents based on the integration of official and crowdsourced data", Journal of Information and Data Management 8, pp. 67–67, 2017.
[8] S. Kumar and D. Toshniwal, "A data mining framework to analyze road accident data", Journal of Big Data 2, pp. 1–18, 2015.
[9] S. Kumar and D. Toshniwal, "A data mining approach to characterize road accident locations", Journal of Modern Transportation 24, pp. 62–72, 2016.
[10] F. M. N. Ali and A. A. M. Hamed, "Usage Apriori and clustering algorithms in WEKA tools to mining dataset of traffic accidents", Journal of Information and Telecommunication 2, pp. 231–245, 2018.
[11] E. Turunen, "Using GUHA data mining method in analyzing road traffic accidents occurred in the years 2004–2008 in Finland", Data Science and Engineering 2, pp. 224–231, 2017.
[12] J. Abellán, G. López, and J. D. OñA, "Analysis of traffic accident severity using decision rules via decision trees", Expert Systems with Applications 40, pp. 6047–6054, 2015.
[13] H. İ. Bülbül, T. Kaya, and Y. Tulgar, "Analysis for status of the road accident occurance and determination of the risk of accident by machine learning in istanbul", In 2016 15th IEEE International Conference on Machine Learning and Applications (ICMLA), IEEE, pp. 426–430, 2016.
[14] S. Gao, R. Pan, R. Yu, and X. Wang, "Research on automated modeling algorithm using association rules for traffic accidents", In 2018 IEEE International Conference on Big Data and Smart Computing (BigComp), IEEE, pp. 127–132, 2018.
[15] Y. Guo, Z. Li, P. Liu, and Y. Wu, "Exploring risk factors with crashes by collision type at freeway diverge areas: Accounting for unobserved heterogeneity", IEEE Access 7 (2019), pp. 11809–11819, 2019.
[16] O. H. Kwon, W. Rhee, and Y. Yoon, "Application of classification algorithms for analysis of road safety risk factor dependencies". Accident Analysis & Prevention 75 (2015), pp. 1–15, 2015.
[17] Z. Li, X. Guo, and J. Sun. "Analysis and research on the temporal and spatial correlation of traffic accidents and illegal activities", In International Conference on Cloud Computing and Security. Springer, pp. 418–428.
[18] J. Wang and Y. Ohsawa, "Evaluating model of traffic accident rate on urban data". In 2016 Federated Conference on Computer Science and Information Systems (FedCSIS). IEEE, pp. 181–186, 2016.
[19] R. Bhavsar, A. Amin, and L. Zala, "Development of model for road crashes and identification of accident spots", International journal of intelligent transportation systems research 19, pp. 99–111, 2021.
[20] A. Iranitalab and A. Khattak, "Comparison of four statistical and machine learning methods for crash severity prediction". Accident Analysis & Prevention 108, pp. 27–36, 2017.
[21] S. Kumar, P. Tiwari, and K. V. Denis, "Augmenting classifiers performance through clustering: A comparative study on road accident data", International Journal of Information Retrieval Research (IJIRR) 8, pp. 57–68, 2018.
[22] N. Mor, H. Sood, and T. Goyal, "Application of machine learning technique for prediction of road accidents in Haryana-A novel approach", Journal of Intelligent & Fuzzy Systems 38, pp. 6627–6636, 2020.
[23] R. Richard and S. Ray, "A tale of two cities: Analyzing road accidents with big spatial data", In 2017 IEEE International Conference on Big Data (Big Data), IEEE, pp. 3461–3470, 2017.
[24] M. Sangare, S. Gupta, S. Bouzefrane, S. Banerjee, and P. Muhlethaler, Exploring the forecasting approach for road accidents: Analytical measures with hybrid machine learning, Expert Systems with Applications, pp. 113855, 2021.
[25] M. S. Satu, S. Ahamed, F. Hossain, T. Akter, and D. M. Farid, "Mining traffic accident data of N5 national highway in Bangladesh employing decision trees", In 2017 IEEE Region 10 Humanitarian Technology Conference (R10-HTC), IEEE, pp. 722–725, 2017.
[26] D. Singh and C. K. Mohan, "Deep spatio-temporal representation for detection of road accidents using stacked autoencoder", IEEE Transactions on Intelligent Transportation Systems 20, pp. 879–887, 2018.
[27] T. Tambouratzis, D. Souliou, M. Chalikias, and A. Gregoriades, "Combining probabilistic neural networks and decision trees for maximally accurate and efficient accident prediction", In The 2010 International Joint Conference on Neural Networks (IJCNN), IEEE, pp. 1–8, 2010.
[28] A. Katsoukis, L. Iliadis, A. Konguetsof, and B. Papadopoulos, "Classification Of Road Accidents Using Fuzzy Techniques", In 2018 Innovations in Intelligent Systems and Applications (INISTA), IEEE, pp. 1–5, 2018.
[29] B. Ryder, B. Gahr, P. Egolf, A. Dahlinger, and F. Wortmann, Preventing traffic accidents with in-vehicle decision support systems-The impact of accident hotspot warnings on driver behaviour, Decision support systems 99, pp. 64–74, 2017.
[30] H. Ren, Y. Song, J. Wang, Y. Hu, and J. Lei, "A deep learning approach to the citywide traffic accident risk prediction", 21st International Conference on Intelligent Transportation Systems (ITSC), IEEE, pp. 3346–3351.
[31] C. Zhang, Y. Shu, and L. Yan. "A Novel Identification Model for Road Traffic Accident Black Spots: A Case Study in Ningbo", China, IEEE Access 7, pp. 140197–140205, 2019.

[32] L. Yu, B. Du, X. Hu, L. Sun, W. Lv, and R. Huang, "Traffic Accident Prediction Based on Deep Spatio-Temporal Analysis", IEEE SmartWorld, Ubiquitous Intelligence & Computing, Advanced & Trusted Computing, Scalable Computing & Communications, Cloud & Big Data Computing, Internet of People and Smart City Innovation (SmartWorld/SCALCOM/UIC/ATC/CBDCom/IOP/SCI). IEEE, pp. 995–1002, 2019.
[33] P. Chapman, et al. Crisp-dm 1.0 Step-by-step data mining guide, 2000.
[34] D. N. Politis, J. P. Romano, M. Wolf, Subsampling, Springer, 1999.
[35] S. P. Vasconcelos, "Source code and corpus", Available at: https://colab.research.google.com/drive/1bSWqT7AyHK5ZaprG7m_kf0fPkhcypPN_, Accessed in: 2023-03-10.
[36] S. P. Vasconcelos, "BERTimbau source code and corpus", Available at: https://colab.research.google.com/drive/1h7RziptpLvCBSaJpiiq3ZbVH2N7Q6n33, Accessed in 2023-03-10.
[37] D. Vaskevitch, Client/server strategies, IDG Books Worldwide, 1995.

NodeGIS: A Container-based Web GIS Application Development Tool

Mateus Queiroz Cunha
Information Systems Laboratory
Federal University of Campina Grande
Campina Grande, Brazil
email: mateusqueiroz@copin.ufcg.edu.br

Cláudio de Souza Baptista
Information Systems Laboratory
Federal University of Campina Grande
Campina Grande, Brazil
email: baptista@computacao.ufcg.edu.br

***Abstract*—Nowadays, several tools which foster the development of Web Geographical Information Systems (GIS) have emerged, associated with the outspread of spatial information. Although there are many solutions, most of them are complex, requiring specific skills from users. Therefore, there exists a demand for a solution that simplifies the process of developing and deploying Web GIS applications. This paper presents NodeGIS, an open-source tool that provides a graphical interface for developing Web GIS applications without code writing or complex server configuration. NodeGIS enables the user to plot vector maps, to perform overlay and customization operations, zooming, panning, tooltip, and spatial and non-spatial queries. NodeGIS uses a container-based and Representational State Transfer (REST) architecture, thus facilitating the deployment of Web GIS applications. NodeGIS can also be used to teach GIS, requiring no software installation from students.**

Keywords*—*GIS; Web GIS; containerization; REST.

I. INTRODUCTION

Along with the ubiquity of spatial information in current applications, several commercial and free tools have emerged, as well as Application Programming Interfaces (APIs) that promote the development of Web Geographic Information System (GIS) applications. Large companies in the GIS sector have provided Web GIS solutions, such as Microsoft Location Technologies - Azure Maps [1], Nokia Here SDK [2], Amazon AWS Location API [3], Google Maps API [4], deck.gl [5], Apple Maps API [6], mapbox [7], Uber API [8], ESRI ArcGIS API [9], QGIS [10], MapServer [11], GeoServer [12], among many other solutions. Also, many database management systems provide native support for spatial data, such as the relational: PostgreSQL/PostGIS, MySQL, SQL Server, Oracle and IBM DB2; NewSQL: SAP Hana, CockRoachDB and SingleStore; and NoSQL: MongoDB, CouchDB, Cassandra, among others.

Concerning desktop GIS, there are many commercial and free solutions, such as QGIS and gvSIG [13]. However, when we look at Web GIS solutions [14] there are numerous solutions using map servers, spatial database servers and frontends. These three tiers contain different complexity and requirements, many times requiring specific skills from the user in the different technologies [15]–[19]. A server-only application that requires specialized frontend development to consume and display data, a severe learning curve before a user can create a useful application, and high prices for commercial (and even educational) purposes are a few examples of encountered challenges. Among Web GIS well known solutions, we mention GeoServer, MapServer, QGIS Server [20], and ArcGIS Server. Therefore, there is a need for a tool that simplifies the process of developing and publishing a Web GIS application [21]. That is the main purpose of the NodeGIS tool.

NodeGIS is an open-source tool that provides a graphical interface for developing a Web GIS without the need to develop code or configure servers. The NodeGIS application deployment is done via Docker [22] containers using two images, one for the frontend and the other one for the backend, in addition to a PostgreSQL/PostGIS container, making the application ready to use with geographic data.

Our contribution concerns to bring the low-effort experience of developing desktop GIS applications to the Web environment. In addition, the NodeGIS tool may be used in the teaching and learning process of geoprocessing techniques, providing both the teacher and students with a tool that allows them to graphically and interactively explore various GIS resources, without needing to install software, using only a browser with an internet connection.

In the remainder of this paper, Section II describes some of the related work, followed by Section III, which presents the architecture and main features of NodeGIS. Section IV describes how to create a Web GIS using NodeGIS. Later, Section V focuses on a user evaluation of the developed solution. Final considerations are the subject of Section VI.

II. RELATED WORK

The evolution of Web Mapping applications in recent years has highlighted the demand for geographic applications that fulfill user requirements [23]. GIS applications have been impacted by the technological advancements of the Web and distributed systems, however, there is a technological gap between Web-oriented information systems and existing GIS solutions [24].

Recent research has focused on the importance of the user interface when working with GIS applications [25]. There are also research on GIS teaching dealing with the positive impact of bringing a desktop GIS experience to a Web GIS application [26][27]. This can also help non-expert users to be able to create Web GIS applications for their respective areas of expertise.

More recently, several Web GIS solutions have been proposed specifically for a single domain, such as landslide and flood mapping [28], spatial accessibility of urban medical facilities [29], seismic vulnerability of old urban areas [30], marine spatial planning [31], and promotion of archaeological and environmental tourism [32]. These research works addressed domain-specific challenges associated with the complexity of using GIS tools.

III. The NodeGIS Solution

Among the main features of NodeGIS, we highlight the addition of vector layers, conventional and spatial queries; customization of map layers; construction of thematic maps; tooltip; interactive query on the map, where it is possible to carry out complex spatial queries in an easy way through the application's interface, with no need for SQL knowledge; search for features on the map; use of multiple spatial databases; data tables, as well as filters and selections that can be applied to the data and visualized on the map.

Figure 1 presents an example of a NodeGIS running application that contemplates the following layers: map of the municipalities, railways and highways of the Brazilian state of Paraíba. In addition, Figure 1 also depicts a spatial query with a buffer operation of the central railway in the state, highlighted in green.

The architectural design of NodeGIS is based on a Representational State Transfer (REST) architecture, segmented into the frontend and backend, associated with one or more spatial relational databases, as detailed in Figure 2. We detail each architectural aspect of the NodeGIS tool in the following subsections.

A. Frontend

The NodeGIS frontend consists of a React [33] application, which uses an implementation of the Flux [34] frontend data flow pattern with four main elements: the View, represented by the React components that display the application data to the user; the Store, responsible for obtaining, manipulating and storing data in the application; the Dispatcher, responsible for managing the data flow, distributing actions coming from the View to their respective Store(s); and, finally, the Action, which are functions that carry data to the Stores derived from user actions. Figure 3 shows the details of the frontend architecture and its components communication.

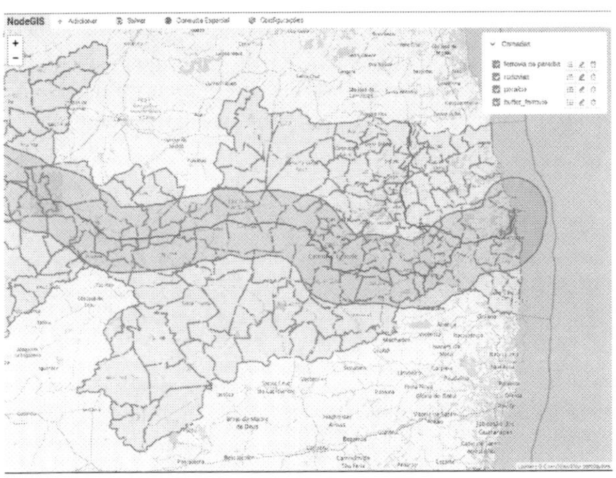

Figure 1. An application example deployed in NodeGIS.

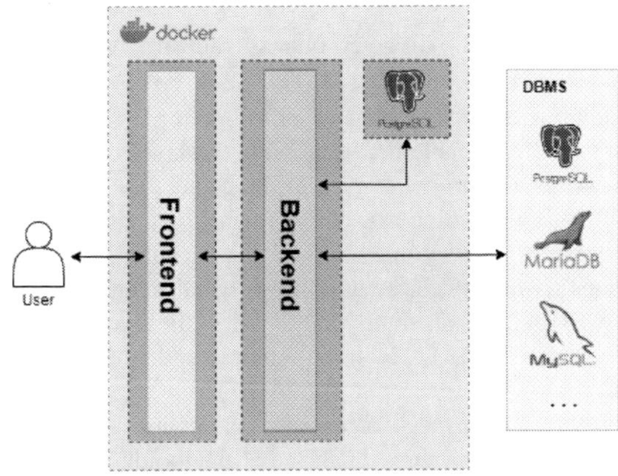

Figure 2. NodeGIS architecture.

The main element of the frontend is the map, developed using the Leaflet [35] library, which has a specific version for React applications. All the data on the map comes from their respective Store, which stores data obtained using HTTP requests to the backend. The backend is responsible for fetching geographic data from the database and converting it to GeoJSON, a geospatial data format that uses JavaScript Object Notation (JSON). The Leaflet library can read the data in a layer on the map directly from GeoJSON format, not requiring any adjustment or additional structuring of the geographic data.

When users want to add a new vector layer to the map, NodeGIS displays a list of all database tables containing spatial data, as well as the names of the spatial columns for each table. Users can also define the data displayed in the tooltip of map features, layer styling, and the layer name on the map using the frontend interface.

A JavaScript object represents each layer, which are stored

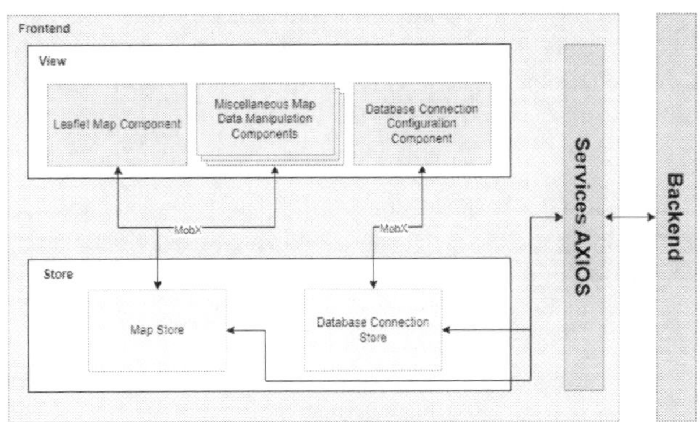

Figure 3. NodeGIS frontend architecture.

in the Map Store. All layer objects follow a well-defined pattern, having the following attributes:

- *name*: name given to the layer.
- *key*: database table that represents the layer.
- *geometryColumn*: data type of the geometry column used in the layer, which can be Polygon, LineString or Point.
- *styles*: an object containing the layer's static style definitions (if the styleType attribute is equal to static).
- *styleType*: an attribute that defines the type of layer styling, whether it is static (static) or thematic (choropleth).
- *displayColumns*: list of objects containing columns that will have their values displayed at the tooltip of the layer features on the map.
- *choroplethStyleDfinition*: an object containing style definitions for thematic maps (if the styleType is equal to choropleth).
- *data*: GeoJSON data returned by querying the database and received as a response to the HTTP request made to the backend.

Leaflet React builds each component using the data from the Map Store, rendering each element on the map. The component called MapContainer defines the external context of the map used in the application and renders the other components corresponding to each element added to the map. Each rendered element uses the data attribute of a vector layer object (described above), sending it to the GeoJSON component. However, we used the CircleMarker component for Point data to emphasize it and enable styling, given that the default markers, displayed when using the GeoJSON component, cannot be customized. That would cause confusion when displaying multiple layers on the map.

Users can also perform spatial querying, an essential feature for a Web GIS application. It can be done either as native SQL queries or by selecting features on the map. The SQL spatial query redirects the query entered manually by the user to the backend and, later, to the database. As for the query which uses feature selection on the map, it englobes the selection of individual features or entire layers according to the chosen spatial operation.

Spatial queries have their structure initially built on the frontend, using a query definition object. Next, the backend receives that object via an HTTP request. This query is then built and subsequently sent to the database.

B. Backend

The NodeGIS backend consists of a web server capable of responding to HTTP requests, following a REST standard, using NodeJS in association with the express framework. As shown in Figure 4, the NodeGIS backend architecture is composed of four segments:

- *Routes*: consists of an access layer that redirects HTTP requests to their respective controllers.
- *Controller*: this layer deals with requests made to the API and their responses based on the data coming from the repository after processing.
- *Repositories*: responsible for building spatial and general queries made to the database.
- *Database Clients*: responsible for the connection to the database(s).

In order to provide versatility and possibility of choice to the user, NodeGIS has compatibility with different spatial relational Database Management Systems (DBMS), namely: PostgreSQL, MySQL, MariaDB, SQLite (with its SpatiaLite extension) and CockroachDB. We implemented a specific database client for each supported DBMS, extending the existing NodeJS database client and enabling NodeGIS to deal with any divergences in SQL implementations. In addition to the external databases used with the application, the backend has also an instance of SQLite running internally. This instance assists in managing the application data, storing which layers the users recorded in their visualization, and storing connection data to the other databases used.

Most of the repositories defined in the backend redirect API calls to the database using basic queries, except for the queryRepository, that is responsible for building the spatial query requests from the frontend. The supported spatial operations are: union, difference, intersection, contains, crosses, touches, disjunction, intersects, buffer, centroid, area, distance, length and perimeter.

When building the spatial query, the frontend makes an HTTP request to the backend containing a body in the format shown in Figure 5. This object comprises the following attributes: a data parameter of the spatial query; the second data parameter of the spatial query; and operation, a string representing the spatial operation to be applied. The second property can be empty for queries that do not apply an operation between different features (or sets of features). For example, that is the case for operations such as buffer and area.

In more detail, the *first* and *second* properties store the tables used in the query as keys of the property object. Each table key also stores properties, which are the *data* and the *geometryColumn*. The *data* property holds a list of unique identifiers of the table features, called Global Identifiers (GIDs). The query uses features whose GID values are in the list — an empty list means considering the entire table. The *geometryColumn* property stores which spatial column of the

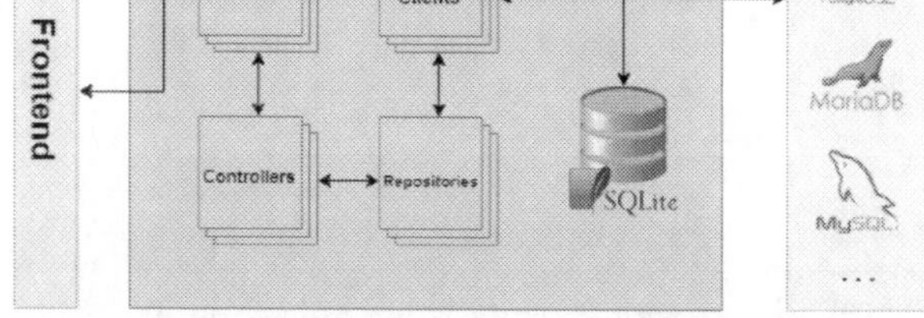

Figure 4. Backend architecture details.

table will be used in the spatial query.

For each of the spatial query parameters, a subquery is built to obtain the requested data, being simply a union of the requested features (if there are feature GIDs in the data property) or the entire table (if the data property is empty). Then, from the subqueries, the spatial function is applied according to the string referenced in the operation property, using the spatial columns of each subquery as parameters for the spatial function.

At the end of the build process, the *ST_AsGeoJSON* function wraps the query to convert the geometry data into the GeoJSON format, making it easier to manipulate, read and represent the data on the map.

C. Containerization

In order to facilitate the Web GIS application deployment, Docker images are available containing the required modules to run the application (frontend and backend). The images are publicly accessible on DockerHub [36].

To facilitate the execution of the Docker images into containers, therefore, the application deployment, two scripts were written in the root directory of the application GitHub repository [37]. One of the scripts runs the application in a local environment (*run-application-local.sh*), and the other in a production environment with public access (*run-application-prod.sh*). These scripts have already each container parameters and environment variables correctly configured. However, a script change is possible according to specific infrastructure needs.

In order to provide a ready-to-use environment, we have a third Docker image of a PostgreSQL database with PostGIS containing sample data. This containerized database is configured as the default and added to the application startup settings.

IV. Creating a Web GIS application using NodeGIS

NodeGIS aimed at providing an application startup that reduces the impacts of integration between frontend, backend and databases, as well as influences of the environment where this application will be used (operating system, for example).

```
"first": {
  "paraiba": {
    "data": [152, 161, 19],
    "geometryColumn:": "geom"
  }
},
"second": {
  "paraiba": {
    "data": [],
    "geometryColumn:": "geom"
  }
},
"operation": "contains"
```

Figure 5. JSON format sent to the backend when doing spatial queries.

Containerization is fundamental in this process. Hence, installing Docker is the only prerequisite for running NodeGIS.

A. Initial Configuration

As described in section III-C, shell scripts were made available in the root directory of the application repository to speed up the application initialization. The scripts are *run-application-local.sh* (local environment) and *run-application-prod.sh* (production environment). Both perform this configuration, however, with different parameters and purposes.

When executing the desired script (e.g., using the "*bash run-application-local.sh*" command), three Docker images on the host machine will be used (and fetched from DockerHub, if not existent locally): the frontend, the backend and the PostgreSQL spatial database. Finally, the three containers will be initialized with their respective parameters, showing an output that contains the hash code of each initialized Docker container and the access URL for the application.

The execution of the scripts is similar in both environments. However, when NodeGIS runs with the production script, it depends on the configuration of the host machine so that external users can access it. The environment must have a public IP and ports 8080 and 8081 unlocked. The *stop-application.sh* script is responsible for stopping NodeGIS execution, using the names assigned to the containers in the execution script.

All script parameters may be modified according to the user needs. An example of this scenario is when a user already has a database containing spatial data, where the initialization of the PostgreSQL container becomes optional, or when the ports used in the production environment need to be changed.

With the application fully initialized, the user can use the features of NodeGIS through the top menu. For example, the feature to add vector layers to the map can be found in the "Add" menu, followed by the "Vector Layer" option. Users can select which database table to use and its spatial column. In addition, users can customize the information displayed in the features' tooltip and the layer style.

B. Spatial Querying

Spatial queries are based on map feature selection and can be accessed in the menu using the "Spatial Query" selector, followed by the "From Selection" option. This feature allows the user to select the desired operation and the parameters needed to carry it out (features or layers). Users can execute queries resulting in geometries, such as the buffer operation. Note that users can change the style of the map feature for better visualization. It is also possible to record the result of spatial queries in new tables in the database, thus possibly reusing them in new queries or saving them as the default visualization of NodeGIS.

Hence, users can easily and quickly: initialize NodeGIS, add vector layers to the map, perform spatial queries and save visualizations generated from existing tables or queries obtained from the NodeGIS.

V. NodeGIS User Evaluation

In order to assess user experience with NodeGIS, we interviewed twenty users using a questionnaire. All users have some experience on information system development. We adapted the Post-Study System Usability Questionnaire (PSSUQ) [38] to have only relevant questions according to a suitable NodeGIS scenario.

The final objective of applying this questionnaire is to obtain metrics regarding the satisfaction of different users with the NodeGIS tool. The questionnaire contains eight statements, which users must classify according to their respective experiences when using the tool. Here are the statements used:

1) I have some Geographic Information Systems (GIS) knowledge/experience.
2) The system was simple to use.
3) The system has all the features and capabilities I expected.
4) I was able to complete tasks and scenarios quickly using the system.
5) The system gave me error messages that clearly told me how to solve the issues.
6) The information (such as online help, on-screen messages and other documentation) provided by the system was clear.
7) The organization of information on the screen was clear.
8) I would recommend the system to others.

Each interviewee evaluated each statement with a score from 1 to 7, where 1 means strongly disagree and 7 strongly agree. When starting the evaluation process, users had access to a descriptive video [39] containing the main features of NodeGIS. In addition, two simple activities were passed on to the interviewees: adding a vector layer to the map and performing a spatial query with the added layer. Then, the use of the tool was unrestricted and unsupervised. All users accessed the same NodeGIS deployment and geographic database.

Observing Figure 6, which contains the result of Question 1, we can see that a little more than 50% of the interviewees declared themselves knowledgeable of GIS techniques, which provides us with a good balance of people who have already had contact with other GIS tools and those who have not.

Figure 7 depicts the remainder of the evaluation results. About 75% of the evaluations of each statement had a score greater than or equal to 5, representing an excellent result. In addition, score 7 was most common in the evaluations of each statement. It also appeared tied to scores 6 or 5.

The high occurrence of scores close to 7 reflects the general average of the evaluation obtained with the questionnaire, which was 6.27 (disregarding the first statement, given that it does not represent the user's experience directly with NodeGIS). The closer this general average is to 7, the greater the user satisfaction when using NodeGIS.

VI. Conclusion

The development of Web GIS applications has increased exponentially. This paper presents The NodeGIS Web GIS application development tool that is based on Docker containers

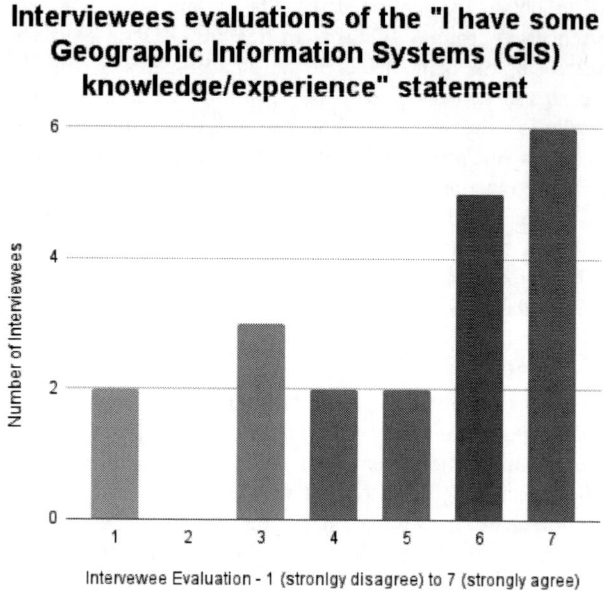

Figure 6. Interviewees evaluation of the "I have some Geographic Information Systems (GIS) knowledge/experience" statement.

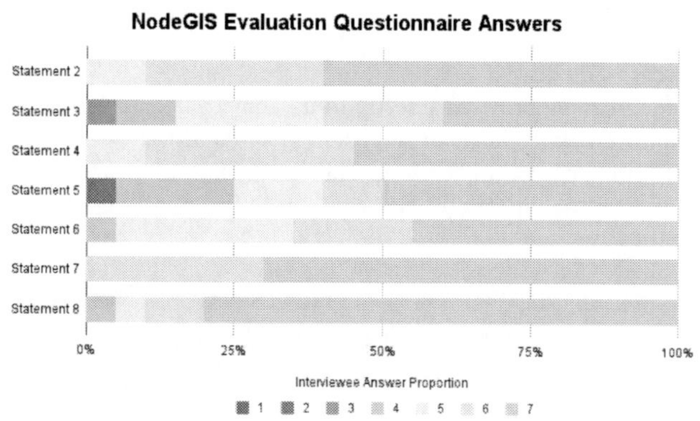

Figure 7. NodeGIS Evaluation Questionnaire Answers.

to simplify deployment. NodeGIS allows plotting vector maps, *overlay* and *layer customization*, *zooming*, *panning*, *tooltip*, performing queries on conventional attributes and various spatial operations on data. NodeGIS also targets aid in teaching GIS without requiring students to install any software.

A NodeGIS user can develop a web GIS application using a simple configuration that requires no specific distributed computation skills. Hence, users may focus on their target domain instead of technical aspects. Even though the NodeGIS tool is designed to be simple, expert users can still use their skills to create unique and advanced spatial queries and use different spatial database management systems.

One of the challenges encountered was the dynamic rendering and data management of the vector layers in the frontend, which can result in a complex implementation when dealing with "pure" JavaScript libraries. The React version of the Leaflet library was essential to overcome this problem, contributing to an implementation that is easy to maintain and evolve.

In future work, we intend to aggregate new simple features to the NodeGIS platform such as table editing, data exportation, and multimedia data association (e.g., associating images and videos with map features). Furthermore, we also plan to include new complex features such as 3D visualizations, point cloud and raster support.

ACKNOWLEDGEMENT

The second author would like to thank the National Council for Scientific and Technological Development (CNPq), Brazil, for partially funding this research.

REFERENCES

[1] "Azure Maps - Geospatial Mapping APIs." Microsoft Azure. https://azure.microsoft.com/en-us/services/azure-maps/ (accessed Mar. 3, 2023).
[2] "Build apps with HERE Maps API and SDK Platform Access." Here Developer. https://developer.here.com/ (accessed Mar. 3, 2023).
[3] "Amazon Location Service." Amazon Web Services. https://aws.amazon.com/location/ (accessed Mar. 3, 2023).
[4] "Google Maps Platform." Google Developers. https://developers.google.com/maps (accessed Mar. 3, 2023).
[5] "deck.gl." deck.gl. https://deck.gl/ (accessed Mar. 3, 2023).
[6] "Apple Maps." Apple Developer. https://developer.apple.com/maps/ (accessed Mar. 3, 2023).
[7] "Maps, geocoding, and navigation APIs & SDKs." Mapbox. https://mapbox.com/ (accessed Mar. 3, 2023).
[8] "Developers." Uber. https://developer.uber.com/ (accessed Mar. 3, 2023).
[9] "ArcGIS Developers." ArcGIS. https://developers.arcgis.com/ (accessed Mar. 3, 2023).
[10] "Welcome to the QGIS project." QGIS. https://www.qgis.org/ (accessed Mar. 3, 2023).
[11] "Welcome to MapServer." MapServer documentation. https://mapserver.org/ (accessed Mar. 3, 2023).
[12] "GeoServer." GeoServer. http://geoserver.org/ (accessed Mar. 3, 2023).
[13] "gvSIG Desktop." Portal gvSIG. http://www.gvsig.com/en/products/gvsig-desktop (accessed Mar. 3, 2023).
[14] D. Moretz, "Internet GIS," in *Encyclopedia of GIS*, Springer International Publishing, 2016, pp. 1–7.
[15] D. Yu and J. Yin, "Internet GIS and system dynamic modeling in urban public safety and security studies: A conceptual framework," in *Lecture Notes in Computer Science*, Springer Berlin Heidelberg, 2011, pp. 207–216.
[16] A. Milosavikevic, "An application framework for rapid development for web based GIS : Ginis Web," in *Geospatial Services and Applications for the Internet*, Springer US, 2008, pp. 48–71.
[17] A. Paiva, E. da Silva, F. Leite, and C. de Souza Baptista, "A multiresolution approach for internet GIS applications," in *Proceedings. 15th International Workshop on Database and Expert Systems Applications, 2004.*, IEEE, 2004, pp. 809–813.
[18] F. Yin and M. Feng, "A webgis framework for vector geospatial data sharing based on open source projects," in *Proceedings. The 2009 International Symposium on Web Information Systems and Applications (WISA 2009)*, Academy Publisher, 2009, pp. 124–127.
[19] E. Nash, P. Korduan, S. Abele, and G. Hobona, "Design requirements for an ajax and web-service based generic internet GIS client," in *11th AGILE International Conference on Geographic Information Science*, University of Girona, Spain, 2008, pp. 1–6.
[20] "QGIS Server GuideManual." QGIS Documentation. https://docs.qgis.org/3.22/en/docs/server_manual/ (accessed Mar. 3, 2023).
[21] T. E. Chow, "The potential of maps APIs for internet GIS applications," *Transactions in GIS*, vol. 12, no. 2, pp. 179–191, Apr. 2008.
[22] "Docker: Accelerated, Containerized Application Development." Docker. https://www.docker.com/ (accessed Mar. 3, 2023).
[23] B. Veenendaal, M. A. Brovelli, and S. Li, "Review of web mapping: Eras, trends and directions," *ISPRS International Journal of Geo-Information*, vol. 6, no. 10, p. 317, Oct. 2017.
[24] A. Rowland, E. Folmer, and W. Beek, "Towards self-service GIS—combining the best of the semantic web and web GIS," *ISPRS International Journal of Geo-Information*, vol. 9, no. 12, p. 753, Dec. 2020.
[25] R. Unrau, A. Kudekar, and C. Kray, "Interaction pattern analysis for scpWebGIS/scp usability evaluation," *Transactions in GIS*, vol. 26, no. 8, pp. 3374–3388, Nov. 2022.
[26] C. Mah, D. Hong, V. Chen, and E. Stefanakis, "First-year engineering students' research experience in web mapping," *Cartographica: The International Journal for Geographic Information and Geovisualization*, vol. 55, no. 1, pp. 53–62, Mar. 2020.
[27] X. Xiang and Y. Liu, "Exploring and enhancing spatial thinking skills: Learning differences of university students within a web-based GIS mapping environment," *British Journal of Educational Technology*, vol. 50, no. 4, pp. 1865–1881, Sep. 2018.
[28] C. Iadanza *et al.*, "IdroGEO: A collaborative web mapping application based on REST API services and open data on landslides and floods in italy," *ISPRS International Journal of Geo-Information*, vol. 10, no. 2, p. 89, Feb. 2021.
[29] J. Wang *et al.*, "Assessing the spatial accessibility of urban medical facilities in multi-level and multi-period scales based on web mapping API and an improved potential model," *ISPRS International Journal of Geo-Information*, vol. 11, no. 11, p. 545, Oct. 2022.
[30] C. Columbro, R. R. Eudave, T. M. Ferreira, P. B. Lourenço, and G. Fabbrocino, "On the use of web mapping platforms to support the seismic vulnerability assessment of old urban areas," *Remote Sensing*, vol. 14, no. 6, p. 1424, Mar. 2022.
[31] A. González, C. Kelly, and A. Rymszewicz, "Advancements in web-mapping tools for land use and marine spatial planning," *Transactions in GIS*, vol. 24, no. 2, pp. 253–267, Dec. 2019.
[32] M. Luppichini *et al.*, "Web mapping and real–virtual itineraries to promote feasible archaeological and environmental tourism in versilia (italy)," *ISPRS International Journal of Geo-Information*, vol. 11, no. 9, p. 460, Aug. 2022.
[33] "React - A JavaScript library for building user interfaces." React. https://reactjs.org/ (accessed Mar. 3, 2023).
[34] "Flux." Flux. https://facebook.github.io/flux/ (accessed Mar. 3, 2023).
[35] "Leaflet - a JavaScript library for interactive maps." Leaflet. https://leafletjs.com/ (accessed Mar. 3, 2023).
[36] M. Cunha, "mateusqc's Profile." Docker Hub. https://hub.docker.com/u/mateusqc (accessed Mar. 3, 2023).
[37] M. Cunha, "mateusqc/node-gis: An application to support the construction of Web GIS." GitHub. https://github.com/mateusqc/node-gis (accessed Mar. 3, 2023).
[38] J. Lewis, "Psychometric evaluation of the post-study system usability questionnaire: The PSSUQ," vol. 2, Jan. 1992, pp. 1259–1263.
[39] M. Cunha, "NodeGIS - Main Features." YouTube. https://youtu.be/SZYesxzrBIc (accessed Mar. 3, 2023).

Advanced Contextualisation Reference Implementation Frameworks in Practice

Coherent Multi-disciplinary Conceptual Knowledge-Spatial Context Discovery Results from the Holocene-prehistoric Volcanological Features and Archaeological Settlement Infrastructure Surveys

Claus-Peter Rückemann
Westfälische Wilhelms-Universität Münster (WWU), Germany;
Unabhängiges Deutsches Institut für Multi-disziplinäre Forschung (DIMF), Germany;
Leibniz Universität Hannover, Germany
Email: ruckema@uni-muenster.de

Abstract—**This paper presents the results of the research on coherent multi-disciplinary contextualisation and symbolic representation of worldwide Holocene-prehistoric volcanological features and discovery of archaeological settlement infrastructures, especially for prehistoric contexts. The research targets flexible context representation, processing, and integration, which includes further development of knowledge resources, visualisation, and chorological and chronological views for analysis, interpretation, decision making, and new insight. The presented practical implementation employs the new Conceptual Knowledge Reference Implementation (CKRI) and the Component Reference Implementations (CRI) framework for conceptual knowledge-based context integration, complements knowledge processing, and geoscientific and spatial processing and visualisation. The goal of this research is the creation of practical knowledge-based methods and tool set components, which provide solid, standardised means for sustainable long-term research. Both, methods and components should enable further continuous development and adoption to future research questions and targets. This paper provides the specific context discovery results, references to all component implementations and realisations. Future research will address further, continuous developments of reference implementations and knowledge resources and the application for advanced scenarios in prehistory and archaeology.**

Keywords–Prehistory; Holocene; Archaeological Settlement Infrastructures; Coherent Multi-disciplinary Conceptual Knowledge Contextualisation with CKRI and CRI Framework.

I. INTRODUCTION

This paper presents the results of the research on practical employment of coherent multi-disciplinary contextualisation and symbolic representation of worldwide Holocene-prehistoric volcanological features and archaeological settlement infrastructures. The goal of contextualisation especially targets prehistoric contexts worldwide, identifying and integrating archaeological, prehistoric objects with objects from other scientific disciplines on equal footing, promoting a coherent multi-disciplinary conceptual methodological approach. The methodological goal of this research is the creation of practical and sustainable knowledge-based methods and tool sets, which provide solid means for sustainable long-term research. The methods and sets of tools should provide standardised components, which can be continuously further developed and adopted to future research questions and targets.

The research targets flexible context processing, integration, and representation, which includes further development of knowledge resources, visualisation, and chorological and chronological views for analysis, interpretation, decision making, and new insight. Two major practical reference implementations were deployed for full implementations, realisations, and continuous further developments: The new versions of the prehistory-protohistory and archaeology Conceptual Knowledge Reference Implementation (CKRI) [1] and the Component Reference Implementations (CRI) framework [2] for conceptual knowledge-based context integration, complements processing, and geoscientific visualisation. CKRI provides the knowledge framework, including multi-disciplinary contexts of natural sciences and humanities [3]. CRI provides the required component groups and components for the implementation and realisation of all the procedural modules.

The reference implementations are based on the fundamental methodology of knowledge complements [4], considering that many facets of knowledge, including prehistory, need to be continuously acquired and reviewed [5]. Creating contextualisation requires to coherently integrate multi-disciplinary knowledge and to enable symbolic representations. Realisations need to integrate a wide range of components as required from participating disciplines, e.g., for dynamical processing, geoprocessing, spatial contextualisation.

The rest of this paper is organised as follows. Sections II and III present the implementations of and conceptual knowledge and the respective components with all required references. Section IV provides the results to selected scenarios. Sections V and VI discuss case scenarios and summarise lessons learned, conclusions, and future work.

II. CONCEPTUAL KNOWLEDGE IMPLEMENTATION

Implementations and realisations are based on the CKRI reference implementation [6], and respective contextualisation. References are capable to integrate required context. Besides the core scope of this knowledge-focussed research on prehistoric, archaeological, and geoscientific questions, procedural complements are employed and extended via the CRI frame reference implementations [7]. Both provide sustainable fundaments for highest levels of reproducibility and standardisation.

Many aspects of knowledge [8], including meaning, can be described using knowledge complements supporting a modern definition of knowledge [9] and subsequent component instrumentation, e.g., considering factual, conceptual, procedural, metacognitive, and structural knowledge. Complements are a means of understanding and targeting new insight, e.g., enabling advanced contextualisation, integration, analysis, synthesis, innovation, prospection, and documentation. Regarding knowledge, it should be taken for granted, that scientific members of any disciplines nowerdays continuously practice and train themselves in development and practical employment of methods, algorithms, and components as required by their disciplines and keep track with how to integrate methods.

A. Coherent conceptual knowledge implementation

Universally coherent multi-disciplinary conceptual knowledge is implemented via the CKRI [6], demonstrated with Universal Decimal Classification (UDC) [10] code references, spanning the main tables based on science and knowledge organisation [11], as shown in Table I.

TABLE I. CKRI IMPLEMENTATION OF COHERENT CONCEPTUAL KNOWLEDGE CONTEXTUALISATION; MAIN TABLES (EXCERPT).

Code / Sign Ref.	*Verbal Description (EN)*
UDC:0	Science and Knowledge. Organization. Computer Science. Information. Documentation. Librarianship. Institutions. ...
UDC:1	Philosophy. Psychology
UDC:2	Religion. Theology
UDC:3	Social Sciences
UDC:5	Mathematics. Natural Sciences
UDC:55	Earth Sciences. Geological sciences
UDC:551.21	Vulcanicity. Vulcanism. Volcanoes. Eruptive phenomena. Eruptions
UDC:551.2...	Fumaroles. Solfataras. Geysers. Hot springs. Mofettes. Carbon dioxide vents. Soffioni
UDC:6	Applied Sciences. Medicine, Technology
UDC:692	Structural parts and elements of buildings
UDC:7	The Arts. Entertainment. Sport
UDC:711	Principles and practice of physical planning. Regional, town and country planning
UDC:8	Linguistics. Literature
UDC:9	Geography. Biography. History
UDC:902	Archaeology
UDC:903	Prehistory. Prehistoric remains, artefacts, antiquities
UDC:904	Cultural remains of historical times

The CKRI is provided in development stage editions, prehistory-protohistory and archaeology E.0.4.8, natural sciences E.0.4.0.

B. Implementation of auxiliaries and operations

Tables II and III show CKRI excerpts of auxiliary tables and signs.

TABLE II. CKRI IMPLEMENTATION OF COHERENT CONCEPTUAL KNOWLEDGE CONTEXTUALISATION; AUXILIARY TABLES (EXCERPT).

Code / Sign Ref.	*Verbal Description (EN)*
UDC (1/9)	Common auxiliaries of place
UDC:(23)	Above sea level. Surface relief. Above ground generally. Mountains
UDC:(24)	Below sea level. Underground. Subterranean
UDC:"..."	Common auxiliaries of time.
UDC:"6"	Geological, archaeological and cultural time divisions
UDC:"62"	Cenozoic (Cainozoic). Neozoic (70 MYBP - present)
UDC:"63"	Archaeological, prehistoric, protohistoric periods and ages

TABLE III. CKRI OPERATION SIGNS EXCERPT, INTEGRATING UDC COMMON AUXILIARY SIGNS (ENGLISH COMMENTS VERSION).

Operation	*Symbol*	
Coordination. Addition	+	(plus sign)
Consecutive extension	/	(oblique stroke sign)
Simple relation	:	(colon sign)
Order-fixing	::	(double colon sign)
Subgrouping	[]	(square brackets)
Introduces non-UDC notation	*	(asterisk)
Direct alphabetical specification.	A/Z	(alphabetic characters)
[Reference listing, itemisation]	;	(semicolon)
[Reference listing, sub-itemisation]	,	(comma)

Consistent multi-disciplinary conceptual knowledge is demonstrated via UDC code references spanning auxiliary tables [11]. Standardised operations (Table III) are employed for creation of reference listings and faceted knowledge, integrating UDC auxiliary signs [11]. Conceptual knowledge in focus can be employed to provide references and facets to any universal knowledge context.

III. COMPONENT IMPLEMENTATIONS

A. Resulting methodological component integration

Integration components, reflecting standards and sustainable modules are based on the major groups of the CRI. The CRI framework is provided in development stage edition E.0.3.9. The ten major CRI component groups are:

1) Conceptual knowledge frameworks.
2) Conceptual knowledge base.
3) Integration of scientific reference frameworks.
4) Formalisation.
5) Methodologies and workflows integration.
6) Prehistory Knowledge Resources.
7) Natural Sciences Knowledge Resources.
8) Inherent representation groups.
9) Scientific context parametrisation.
10) Structures and symbolic representation.

All parts were realised based on CRI components, with realisations fully referenced in the following sections.

The conceptual knowledge was realised for all disciplines via the CKRI conceptual knowledge framework [6] and operations (Table III). CKRI is demonstrated with UDC [10] references. For demonstration, CKRI references are illustrated via the multi-lingual UDC summary [10] released by the UDC Consortium, Creative Commons license [12].

Relevant scientific practices, frameworks, and standards from disciplines and contexts are integrated with the Knowledge Resources, e.g., here details regarding volcanological features, chronologies, spatial information, and Volcanic Explosivity Index (VEI) [13]. Corresponding coherent complementary results and details on faceting are available for a whole inventory of volcanological features groups [1].

All integration components, for all disciplines, require an explicit and continuous formalisation [14] process. The formalisation includes computation model support, e.g., parallelisation standards, OpenMP [15], Reg Exp patterns, e.g., Perl Compatible Regular Expressions (PCRE) [16]. Here, common scale of entities for primary objects is 10^3 and for secondary objects 10^4–10^5. Processing operations [17] were parallelised for primary (n_1) features groups with respective instances.

Methodologies for creating and utilising methods include model processing, remote sensing, spatial mapping, high information densities, and visualisation. Respective contextualisation for scenarios in prehistory should be done under conditions especially reflecting state-of-the-art methods, e.g., spatial operations, triangulation, gradient computation, and projection.

The symbolic representation of the contextualisation can be done with a wide range of methods, algorithms, and available components, e.g., implemented here via LX Professional Scientific Content-Context-Suite (LX PSCC Suite) deploying the Generic Mapping Tools (GMT) [17] for visualisation.

Prehistoric objects and contexts are taken from The Prehistory and Archaeology Knowledge Archive (PAKA), which is in continuous development for more than three decades [18], released by DIMF [19].

Several coherent systems of major natural sciences' context object groups from KR realisations have been implemented, especially Knowledge Resources (KR) focussing on volcanological features [13] deployed with in depth contextualisation and with a wide range [10] of contexts [20] and structures [21].

The contextualisation solution can employ state-of-the-art results from many disciplines, e.g., context from the natural sciences resources, integrating their inherent representation and common utilisation, e.g., points, polygons, lines, and spatial techniques and standards. Here, resources are Digital Elevation Models (DEM), High Resolution (HR) (Space) Shuttle Radar Topography Mission (SRTM) [22] data fusion [23], HR Digital Chart of the World (DCW) [24], and Global Self-consistent Hierarchical High-resolution Geography (GSHHG) [25].

Scientific context parametrisation of prehistoric targets can use the overall insight from all disciplines, e.g., parametrising algorithms and creating palaeolandscapes.

Structure is an organisation of interrelated entities in a material or non-material object or system [21]. Here, relevant examples of sustainable implementations are NetCDF [26] based standards, including advanced features, hybrid structure integration, and parallel computing support (PnetCDF).

Overall, all parts of the solution were implemented and realised via these components. Especially, GMT modules were deployed for select procedures together with PCRE and Perl filters. Spatial distance dependencies of objects and conditional decision criteria were realised via GMT geodesic calculation, which is very accurate using the Vincenty algorithm [27].

IV. Scenarios, Implementations and Results

The results for a multi-disciplinary case scenario from the current research with two primary case instances were choosen, Holocene-prehistoric volcanological features of strato volcanoes and maars (CKRI: UDC:511.2...), with geospherical calculations on a global scale and context discovery with coherently classified archaeological settlement infrastructure instances (CKRI: UDC:711....,692,903,902,...) in geospherical radii of 300 km spatial distance from primary objects.

A. Methodological approach

The method can be summarised as follows.

- Selection of KR, components, primary and secondary object types, symbolic representation, ...
- Conceptual knowledge assignment.
- Selection of chronological properties.
- Selection of primary objects.
- Selection of secondary objects.
- Calculation of secondary objects' geospherical spatial distances.
- Parallelisation of conceptual knowledge processing.
- Parametrisation of symbolic representation.
- Parallelisation of context data processing.
- Visualisation processing.
- (Further development of resources and implementations by the specific disciplines.)

In new applications, all steps and items should be carefully and intentionally addressed for any intelligent employment, depending on the research questions and contexts.

B. Resulting context groups

An excerpt of the two primary context groups and criteria (@ and ⓑ) and contextualisation of archaeological settlement infrastructures is shown in (Table IV).

TABLE IV. SCENARIO CONTEXT GROUPS AND CRITERIA: VOLC. FEATURES / ARCH. SETTLEMENT INFRASTRUCTURES (EXCERPT).

Context n_1		*Context n_2*
Geosciences		Archaeology / prehistory
Geoscientific features objects		Prehistoric object groups
Volc. features groups		Settlement infrastructures
• **Strato volcano**	@ ⇒	• Viereckschanze
• **Maar**	ⓑ ⇒	• Dwelling
• Complex volcano		• Long house
• Explosion crater		• Midden context
• Shield volcano		• Farm hut
• Subglacial volcano		• Enclosure
• Submarine volcano		• Roundhouse
• Volcanic field		• Siedlungsplatz
• Fissure vent		• Homestead
• Cone		• Hut circle
• Dome		• [individually named]
• ...		• ...
[Type Instances ...]		[Type instances ...]
Decision Criteria (n_1)		*Decision Criteria (n_2)*
Conceptual context (CKRI)		Conceptual context (CKRI)
Feature object type		Prehistoric object type
Chronology conditions		Chronology conditions
Chorology / positional conditions		Chorology / positional conditions
Object attributes		Object attributes
...		Geospherical spatial distance
...		(n_1-instance-conditional)
...		Parametrisation, ...,
...		Calculation / analysis

The primary decision criteria include conceptual context, feature object type, chronology, and position. The secondary decision criteria include conceptual context, prehistoric object type, chronology, position, and conditional geospherical spatial distance depending on respective primary objects. Prehistoric object groups include all available language representations, e.g., 'en' and 'de'. Here, the first primary object group defines the spatial projection for consecutive primary groups.

C. Resulting context discovery matrices

Table V shows an excerpt of the result matrix of Holocene-prehistoric volcanological features groups and respective facets, namely conceptual knowledge, chronology, and chorology for the two scenarios (@ and ⓑ in Table IV). The result matrix includes conceptual knowledge view groups [10] based on CKRI references [6], factual knowledge from the Knowledge Resources objects, and respective country codes. Context example references for the features groups facets show Prehistoric Volcanic Activity (PVA), Historic Volcanic Activity (HVA), and Continued Volcanic Activity (CVA), e.g., latent volcanic activity. PVA are consequence of the Holocene-prehistoric chronological contextualisation for all objects in the resulting volcanological features groups. Cases for which further facts are holding true can also allow past-prehistoric contextualisation, e.g., with HVA and CVA. Resulting context discovery matrices for both primary case instances of the scenario for multi-disciplinary contextualisation of settlements are given in Tables VI and VII. Instances @ and ⓑ refer to Table IV. Figure 1 shows a corresponding visualisation of the calculation results of the context discovery for both instances.

TABLE V. RESULT MATRIX OF HOLOCENE-PREHISTORIC VOLCANOLOGICAL FEATURES GROUPS FACETS (EXCERPT, (α), (β)). IT INCLUDES CONCEPTUAL KNOWLEDGE VIEW GROUPS [10] (CKRI), VOLCANIC ACTIVITY, CONTEXTS, KNOWLEDGE RESOURCES OBJECTS, AND COUNTRY CODES (EXCERPT).

Multi-disciplinary Conceptual Knowledge Facets	*Chronology Facets*		*Chorology Facets*		
Volcanological Features Conceptual Knowledge View / Facets Group	*Volcanic Activity*	*Context*	*KR Object & Ref.*		*Country Code*
CKRI: UDC:551.21,550.3,(23),STRATO_VOLCANO;"62"...	Holocene	PVA / HVA	Agua de Pau	①	PT
CKRI: UDC:551.21,550.3,(23),STRATO_VOLCANO;"62"...	Holocene	PVA	Alngey	②	RU
CKRI: UDC:551.21,550.3,(23),STRATO_VOLCANO;"62"...	Holocene	PVA / HVA	Azuma	③	JP
CKRI: UDC:551.21,550.3,(23),STRATO_VOLCANO;"62"...	Holocene	PVA / HVA	Hekla	④	IS
CKRI: UDC:551.21,550.3,(23),STRATO_VOLCANO;"62"...	Holocene	PVA	...		...
CKRI: UDC:551.2...,551.21,550.3,(23),...MAARS_FEATURES;"62"...	Holocene	PVA	Cerro Tujle	①	CL
CKRI: UDC:551.2...,551.21,550.3,(23),...MAARS_FEATURES;"62"...	Holocene	PVA / HVA	Suoh	②	ID
CKRI: UDC:551.2...,551.21,550.3,(23),...MAARS_FEATURES;"62"...	Holocene	PVA / HVA	Ukinrek Maars	③	US
CKRI: UDC:551.2...,551.21,550.3,(23),...MAARS_FEATURES;"62"...	Holocene	PVA / (CVA)	West Eifel Volcanic Field	④	DE
CKRI: UDC:551.2...,551.21,550.3,(23),...MAARS_FEATURES;"62"...	Holocene	PVA	...		...
...	...	...	...		...

TABLE VI. RESULTING SETTLEMENT INFRASTRUCTURES FROM CONTEXTUALISATION WITH HOLOCENE-PREHISTORIC STRATO VOLCANO (α) VOLCANOLOGICAL FEATURES GROUP (EXCERPT), INCLUDING CONCEPTUAL KNOWLEDGE VIEW GROUPS [10] (CKRI).

Multi-disciplinary Conceptual Knowledge Facets	*Chronology Facets*		*Chorology Facets*		
Prehistorical Conceptual Knowledge View / Facets Group	*Dependency*	*Context*	*Knowledge Resources Object*	*Count*	*Ref. & Range*
CKRI: UDC:711.....,692,903,902,SETTLEMENT_INFRASTRUCTURE;...	Synchronous	MN-LIA	Grota do Medo	$\sum = 1$	① 300 km
CKRI: UDC:711.....,692,903,902,SETTLEMENT_INFRASTRUCTURE;...	Synchronous	MN-LIA	–	$\sum = 0$	② 300 km
CKRI: UDC:711.....,692,903,902,SETTLEMENT_INFRASTRUCTURE;...	Synchronous	MN-LIA	Sakiyama Kaizuka		③ 300 km
CKRI: UDC:711.....,692,903,902,SETTLEMENT_INFRASTRUCTURE;...	Synchronous	MN-LIA	Togariishi		③ 300 km
CKRI: UDC:711.....,692,903,902,SETTLEMENT_INFRASTRUCTURE;...	Synchronous	MN-LIA	Yaze		③ 300 km
CKRI: UDC:711.....,692,903,902,SETTLEMENT_INFRASTRUCTURE;...	Synchronous	MN-LIA	...	$\sum = 14$	③ 300 km
CKRI: UDC:711.....,692,903,902,SETTLEMENT_INFRASTRUCTURE;...	Synchronous	MN-LIA	Flókatóftir		④ 300 km
CKRI: UDC:711.....,692,903,902,SETTLEMENT_INFRASTRUCTURE;...	Synchronous	MN-LIA	Þjóðveldisbærinn		④ 300 km
CKRI: UDC:711.....,692,903,902,SETTLEMENT_INFRASTRUCTURE;...	Synchronous	MN-LIA	Vogur		④ 300 km
CKRI: UDC:711.....,692,903,902,SETTLEMENT_INFRASTRUCTURE;...	Synchronous	MN-LIA	Stöðvarfjörður	$\sum = 4$	④ 300 km

TABLE VII. RESULTING SETTLEMENT INFRASTRUCTURES FROM CONTEXTUALISATION WITH HOLOCENE-PREHISTORIC MAARS (β) VOLCANOLOGICAL FEATURES GROUP (EXCERPT), INCLUDING CONCEPTUAL KNOWLEDGE VIEW GROUPS [10] (CKRI).

Multi-disciplinary Conceptual Knowledge Facets	*Chronology Facets*		*Chorology Facets*		
Prehistorical Conceptual Knowledge View / Facets Group	*Dependency*	*Context*	*Knowledge Resources Object*	*Count*	*Ref. & Range*
CKRI: UDC:711.....,692,903,902,SETTLEMENT_INFRASTRUCTURE;...	Synchronous	MN-LIA	Potrero de Payogasta		① 300 km
CKRI: UDC:711.....,692,903,902,SETTLEMENT_INFRASTRUCTURE;...	Synchronous	MN-LIA	Pucará de Tilcara		① 300 km
CKRI: UDC:711.....,692,903,902,SETTLEMENT_INFRASTRUCTURE;...	Synchronous	MN-LIA	Tulor		① 300 km
CKRI: UDC:711.....,692,903,902,SETTLEMENT_INFRASTRUCTURE;...	Synchronous	MN-LIA	...	$\sum = 8$	① 300 km
CKRI: UDC:711.....,692,903,902,SETTLEMENT_INFRASTRUCTURE;...	Synchronous	MN-LIA	Segayun megalithic site	$\sum = 1$	② 300 km
CKRI: UDC:711.....,692,903,902,SETTLEMENT_INFRASTRUCTURE;...	Synchronous	MN-LIA	–	$\sum = 0$	③ 300 km
CKRI: UDC:711.....,692,903,902,SETTLEMENT_INFRASTRUCTURE;...	Synchronous	MN-LIA	Federlesmahd VS		④ 300 km
CKRI: UDC:711.....,692,903,902,SETTLEMENT_INFRASTRUCTURE;...	Synchronous	MN-LIA	Gelbrunn Wald VS		④ 300 km
CKRI: UDC:711.....,692,903,902,SETTLEMENT_INFRASTRUCTURE;...	Synchronous	MN-LIA	Hardheim VS		④ 300 km
CKRI: UDC:711.....,692,903,902,SETTLEMENT_INFRASTRUCTURE;...	Synchronous	MN-LIA	...	$\sum = 180$	④ 300 km

V. DISCUSSION OF CASE SCENARIO RESULTS

Implementation and realisation provide a seamlessly coherent multi-disciplinary conceptual knowledge contextualisation for the case scenario and its instances.

The context discovery result matrices (Tables VI and VII) for the instances (α) and (β) both refer to n_2 in Table IV. Especially, these secondary object groups include objects from Middle Neolithic (MN) to at least Late Iron Age (LIA), including ages in between, e.g., Bronze Age. The objects groups comprise all types of settlement infrastructures, e.g., Celtic ramparts, Viereckschanzen (VS), and middens with settlement contexts. The resulting group of strato volcanoes aligns along 0°/360° longitude (Figure 1). An appropriate Transverse Mercator projection was choosen in order to minimise the distortion along a respective meridian for the generation of the primary results of strato volcanoes and results for other consecutive, secondary, contextualised volcanological features, e.g., maars. The CRI framework components were employed for all steps, including knowledge organisation, conceptual and spatial calculation, and visualisation. Primary objects, strato volcano (medium green volcano symbol) and maars (light green volcano symbol) are marked as well as resulting secondary objects, settlements (blue rectangular symbols), all in their precise georeferenced position. Resulting conceptual knowledge is given for these objects. Resulting sums of secondary discovery objects were calculated. Any case can be dynamically contextualised with coherent multi-disciplinary knowledge, as demonstrated for geosciences, prehistory, and archaeology, e.g., referring to prehistoric object properties and

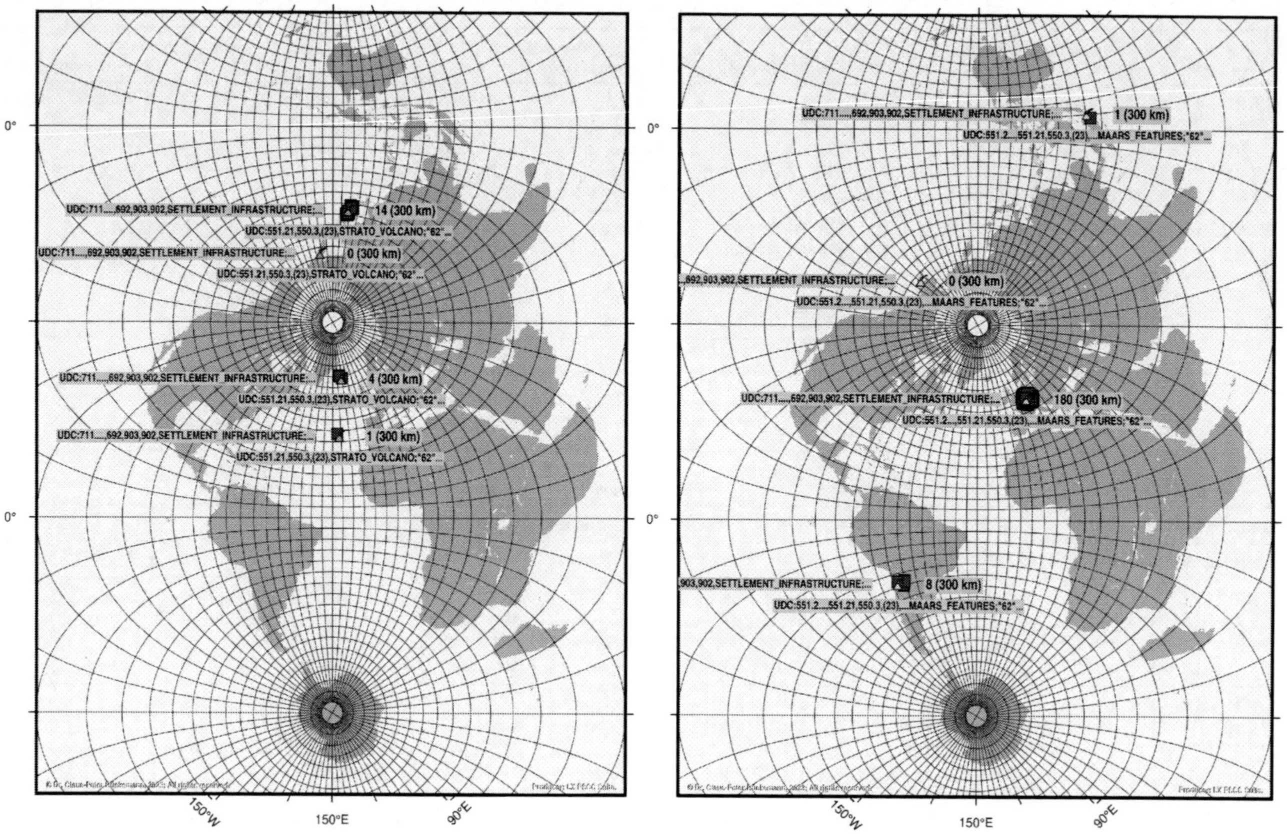

(a) Strato volcano group @: Resulting archaeological settlement infrastructures. (b) Maars group (β): Resulting archaeological settlement infrastructures.

Figure 1. Contextualisation: Holocene-prehistoric volcanological features groups for the two case scenarios and the resulting settlement infrastructures. Coherent multi-disciplinary context integration and results based on CKRI, chronological, and chorological criteria (excerpts, Transverse Mercator projections).

excavation results and targeting new insight from geoscientific and multi-disciplinary context integration.

Any resulting contextualisation matrices and coherent conceptual and faceted knowledge can further be input to consecutive contextualisation processes. The more, solutions with individual methods and workflows can be created for countless different questions and situations.

VI. Conclusion

This research implemented and realised multi-disciplinary contextualisation, based on employing the contextualisation reference implementation frameworks for coherent multi-disciplinary conceptual knowledge-spatial context discovery achieved its goals and proved efficient and sustainable. The case scenarios for context discovery of archaeological settlement infrastructures for Holocene-prehistoric volcanological features resulted in valuable contextualisation potential and possible insight.

The contextualisation integrates conceptual, factual, procedural, structural, and metacognitive knowledge complements. Based on the methodological approach, complements can be identified and assigned during the contextualisation processes. An excerpt of complements relevant for this case scenario are CKRI classification, position data, calculation algorithms, content structures, and parametrisation experiences.

The methods and reference implementations can be efficiently and effectively employed for practical implementations and realisations for multi-disciplinary research, especially in prehistory, archaeology, natural sciences, and humanities. The solutions provide countless facilities and modules for adopting to individual solutions.

Future research will address archaeological settlement infrastructures and further object groups and new models for their coherent multi-disciplinary contextualisation. Prehistoric object groups are matter of future surveys and investigations, e.g., context artefacts and soil characteristics, including further integration and processing of knowledge complements, georeferencing, spatial and satellite data processing.

Acknowledgements

This ongoing research is supported by scientific organisations and individuals. We are grateful to the "Knowledge in Motion" (KiM) long-term project, Unabhängiges Deutsches Institut für Multi-disziplinäre Forschung (DIMF), for partially funding this research, implementation, case studies, and publication under grants D1988FGU90618, D1988FGU90842, D2022F1P05308, and D2022F1P05312. and to its senior scientific members and members of the permanent commission of the science council, especially to Dr. Friedrich Hülsmann, Gottfried Wilhelm Leibniz Bibliothek (GWLB) Hannover, to Dipl.-Biol. Birgit Gersbeck-Schierholz, Leibniz Universität Hannover for fruitful discussion, inspiration, and practical multi-disciplinary contextualisation and case studies. We are grateful to Dipl.-Geogr. Burkhard Hentzschel and Dipl.-Ing. Eckhard Dunkhorst, Minden, Germany, for prolific discussion and exchange of practical spatial, UAV, and context scenarios. We are grateful to Dipl.-Ing. Hans-Günther Müller, Göttingen, Germany, for providing specialised, manufactured high end computation, storage, and visualisation solutions. We are grateful to The Science and High Performance Supercomputing Centre (SHPSC) for long-term support. / DIMF-PIID-DF98_007; URL: https://scienceparagon.de/cpr .

REFERENCES

[1] C.-P. Rückemann, "Faceting the Holocene-prehistoric Inventory of Volcanological Features Groups Towards Sustainable Multi-disciplinary Context Integration in Prehistory and Archaeology Based on the Methodology of Coherent Conceptual Knowledge Contextualisation," International Journal on Advances in Intelligent Systems, vol. 15, no. 3&4, 2022, pp. 115–129, ISSN: 1942-2679, LCCN: 2008212456 (Library of Congress), URL: http://www.iariajournals.org/intelligent_systems [accessed: 2023-03-12].

[2] C.-P. Rückemann, "Component Framework Implementation and Realisation for Development and Deployment of a Coherent Multi-disciplinary Conceptual Knowledge-based Holocene-prehistoric Inventory of Volcanological Features Groups and Faceting," International Journal on Advances in Intelligent Systems, vol. 15, no. 3&4, 2022, pp. 103–114, ISSN: 1942-2679, LCCN: 2008212456 (Library of Congress), URL: http://www.iariajournals.org/intelligent_systems [accessed: 2023-03-12].

[3] C.-P. Rückemann, "The Information Science Paragon: Allow Knowledge to Prevail, from Prehistory to Future – Approaches to Universality, Consistency, and Long-term Sustainability," The International Journal "Information Models and Analyses" (IJ IMA), vol. 9, no. 3, 2020, pp. 203–226, Markov, K. (ed.), ISSN: 1314-6416 (print), Submitted accepted article: November 18, 2020, Publication date: August 17, 2021, URL: http://www.foibg.com/ijima/vol09/ijima09-03-p01.pdf [accessed: 2023-03-12].

[4] C.-P. Rückemann, "From Knowledge and Meaning Towards Knowledge Pattern Matching: Processing and Developing Knowledge Objects Targeting Geoscientific Context and Georeferencing," in Proc. GEOProcessing 2020, November 21–25, 2020, Valencia, Spain, 2020, pp. 36–41, ISSN: 2308-393X, ISBN-13: 978-1-61208-762-7.

[5] R. Gleser, Zu den erkenntnistheoretischen Grundlagen der Prähistorischen Archäologie. Leiden, 2021, (title in English: On the Epistemological Fundaments of Prehistorical Archaeology), in: M. Renger, S.-M. Rothermund, S. Schreiber, and A. Veling (Eds.), Theorie, Archäologie, Reflexion. Kontroversen und Ansätze im deutschsprachigen Diskurs, (in print).

[6] C.-P. Rückemann, "Towards Conceptual Knowledge Reference Implementations for Context Integration and Contextualisation of Prehistory's and Natural Sciences' Multi-disciplinary Contexts," International Journal on Advances in Systems and Measurements, vol. 14, no. 1&2, 2021, pp. 113–124, ISSN: 1942-261x, LCCN: 2008212470 (Library of Congress), URL: http://www.iariajournals.org/systems_and_measurements [accessed: 2023-03-12].

[7] C.-P. Rückemann, "Towards a Component Reference Implementations Frame for Achieving Multi-disciplinary Coherent Conceptual and Chorological Contextualisation in Prehistory and Prehistoric Archaeology," International Journal on Advances in Systems and Measurements, vol. 14, no. 1&2, 2021, pp. 103–112, ISSN: 1942-261x, LCCN: 2008212470 (Library of Congress), URL: http://www.iariajournals.org/systems_and_measurements [accessed: 2023-03-12].

[8] L. W. Anderson and D. R. Krathwohl, Eds., A Taxonomy for Learning, Teaching, and Assessing: A Revision of Bloom's Taxonomy of Educational Objectives. Allyn & Bacon, Boston, MA (Pearson Education Group), USA, 2001, ISBN: 978-0801319037.

[9] C.-P. Rückemann, F. Hülsmann, B. Gersbeck-Schierholz, P. Skurowski, and M. Staniszewski, Knowledge and Computing. Post-Summit Results, Delegates' Summit: Best Practice and Definitions of Knowledge and Computing, Sept. 23, 2015, The Fifth Symp. on Adv. Comp. and Inf. in Natural and Applied Sciences (SACINAS), The 13th Int. Conf. of Num. Analysis and Appl. Math. (ICNAAM), Sept. 23–29, 2015, Rhodes, Greece, 2015, pp. 1–7, DOI: 10.15488/3409.

[10] "Multilingual Universal Decimal Classification Summary," 2012, UDC Consortium, 2012, Web resource, v. 1.1. The Hague: UDC Consortium (UDCC Publication No. 088), URL: http://www.udcc.org/udcsummary/php/index.php [accessed: 2023-03-12].

[11] "UDC Summary Linked Data," 2023, Universal Decimal Classification (UDC), UDC Consortium, URL: https://udcdata.info/ [accessed: 2023-03-12].

[12] "Creative Commons Attribution Share Alike 3.0 license," 2012, URL: http://creativecommons.org/licenses/by-sa/3.0/ [accessed: 2023-03-12], (first release 2009, subsequent update 2012).

[13] C.-P. Rückemann, "Cognostics and Knowledge Used With Dynamical Processing," International Journal on Advances in Software, vol. 8, no. 3&4, 2015, pp. 361–376, ISSN: 1942-2628, LCCN: 2008212462 (Library of Congress), URL: http://www.iariajournals.org/software/ [accessed: 2023-03-12].

[14] C.-P. Rückemann, R. Pavani, B. Gersbeck-Schierholz, A. Tsitsipas, L. Schubert, F. Hülsmann, O. Lau, and M. Hofmeister, Best Practice and Definitions of Formalisation and Formalism. Post-Summit Results, Delegates' Summit: The Ninth Symp. on Adv. Comp. and Inf. in Natural and Applied Sciences (SACINAS), The 17th Int. Conf. of Num. Analysis and Appl. Math. (ICNAAM), Sept. 23–28, 2019, Rhodes, Greece, 2019, DOI: 10.15488/5241.

[15] L. Dagum and R. Menon, "OpenMP: an industry standard API for shared-memory programming," Computational Science & Engineering, (IEEE), vol. 5, no. 1, 1998, pp. 46–55.

[16] "Perl Compatible Regular Expressions (PCRE)," 2023, URL: https://www.pcre.org/ [accessed: 2023-03-12].

[17] P. Wessel, W. H. F. Smith, R. Scharroo, J. Luis, and F. Wobbe, "The Generic Mapping Tools (GMT)," 2020, URL: http://www.generic-mapping-tools.org/ [accessed: 2023-03-12], URL: http://gmt.soest.hawaii.edu/ [accessed: 2023-03-12].

[18] C.-P. Rückemann, "Information Science and Inter-disciplinary Long-term Strategies – Key to Insight, Consistency, and Sustainability: Conceptual Knowledge Reference Methodology Spanning Prehistory, Archaeology, Natural Sciences, and Humanities," International Tutorial, DataSys Congress 2020, Sept. 27 – Oct. 1, 2020, Lisbon, Portugal, 2020, URL: http://www.iaria.org/conferences2020/ProgramINFOCOMP20.html [accessed: 2023-03-12].

[19] "The Prehistory and Archaeology Knowledge Archive (PAKA) license," 2023, (release 2023), Unabhängiges Deutsches Institut für Multidisziplinäre Forschung (DIMF): All rights reserved. Rights retain to the contributing creators.

[20] C.-P. Rückemann, "Prehistory's and Natural Sciences' Multi-disciplinary Contexts: Contextualisation and Context Integration Based on Universal Conceptual Knowledge," in Proc. INFOCOMP 2021, May 30 – June 3, 2021, Valencia, Spain, 2021, ISSN: 2308-3484, ISBN: 978-1-61208-865-5.

[21] C.-P. Rückemann, "The Impact of Information Science Accompanied Structural Information on Computation of Knowledge Pattern Matching and Processing: A Prehistory, Archaeology, Natural Sciences, and Humanities Conceptual Integration Perspective," in Proc. INFOCOMP 2020, Sept. 27 – Oct. 1, 2020, Lisbon, Portugal, 2020, ISBN: 978-1-61208-807-5, URL: http://www.thinkmind.org/index.php?view=article&articleid=infocomp_2020_1_10_60015 [accessed: 2023-03-12].

[22] B. Tozer, D. T. Sandwell, W. H. F. Smith, C. Olson, J. R. Beale, and P. Wessel, "Global Bathymetry and Topography at 15 Arc Sec: SRTM15+," Earth and Space Science, vol. 6, no. 10, Oct. 2019, pp. 1847–1864, ISSN: 2333-5084, DOI: 10.1029/2019EA000658.

[23] C. L. Olson, J. J. Becker, and D. T. Sandwell, "SRTM15_PLUS: Data fusion of Shuttle Radar Topography Mission (SRTM) land topography with measured and estimated seafloor topography," (NCEI Accession 0150537), National Centers for Environmental Information (NCEI), NOAA, 2016.

[24] P. Wessel, "DCW for GMT 6 or later," 2022, URL: http://www.soest.hawaii.edu/pwessel/dcw/ [accessed: 2023-03-12].

[25] P. Wessel, "GSHHG," 2017, URL: http://www.soest.hawaii.edu/pwessel/gshhg/ [accessed: 2023-03-12].

[26] "Network Common Data Form (NetCDF)," 2021, DOI: 10.5065/D6H70CW6, URL: http://www.unidata.ucar.edu/software/netcdf/ [accessed: 2023-03-12].

[27] T. Vincenty, "Direct and inverse solutions of geodesics on the ellipsoid with application of nested equations," Surv. Rev., vol. XXII, no. 176, 1975, pp. 88–93.

Evaluating the Quality of Authoritative Linked Data Models

Hanan Muhajab
School of Computer Science and Informatics,
Cardiff University, Cardiff, UK, and
Jazan University, Saudi Arabia
email: muhajabh@cardiff.ac.uk

Alia Abdelmoty
School of Computer Science and Informatics,
Cardiff University
Cardiff, UK
email: abdelmotyai@cardiff.ac.uk

***Abstract*—Currently, there is a shortage of studies focusing on analyzing existing authoritative geographic ontologies to promote their reuse. This work attempts to fill this gap by reviewing and evaluating four authoritative geographic ontologies on the web. Evaluation is carried out using a set of quantitative quality metrics. Results provide insight into the accuracy, complexity, and completeness of the ontologies and highlight the need for further studies in the heterogeneity of their underlying models.**

***Index Terms*—Geographic Ontology; Linked Data; Metrics.**

I. INTRODUCTION

Semantic Web and Linked Data technologies have been considered for the representation and sharing of authoritative geographic data sets. For example, the Ordnance Survey, the mapping agency of Great Britain [1], has five defined ontologies, and provide open data sets of approximately 64,342,201 triples. Similarly, several spatial linked data sets for the Netherlands, Norway, Germany, Ireland, and Spain are published, and presented at the Knowledge Graph in Action conference (KGiA) [2]. The different providers propose different ontologies for the representation of their data. The heterogeneity of the ontologies is a limitation of their reuse. Evaluation of an ontology refers to measuring its quality to determine its fitness for purpose. The evaluation process involves two perspectives: the provider's perspective, that focus on the accuracy of presentation, error, and quality of naming, and the consumer's perspective, which focuses on the data level and ease of understanding the model. Several studies have been conducted to assess authoritative geographical linked data from the developer's perspective. For example, Debattista et al. [3] evaluated Ordnance Survey Ireland (OSI) using the Luzzu and OOPS platforms [4]. There is a need to assess the authoritative geographic ontologies from the user's perspective to better facilitate their understanding and reuse, as recommended in KGiA [2]. This work analyses a sample of established authoritative geographic ontologies available on the web to examine the complexity of their representation and their completeness of representation. Results show how spatial completeness is limited in most of the studied ontologies and thus further work needs to examine this factor in the future.

The remainder of this paper is organized as follows: section II explains the methodology used in the study. The evaluation results are discussed in section III, and the paper is concluded in section IV.

II. METHOD

Four ontologies were downloaded. These are the administrative units for the UK (O_1), Ireland (O_2), Greece (O_3), and France (O_4), as shown in Table I. A metric-based evaluation method was used as it provides a quantitative and objective way of comparison. Figure 1 shows the metrics used in the evaluation process. The schema metrics, graph metrics, and knowledge base metrics are derived from OntoMetrics [5].

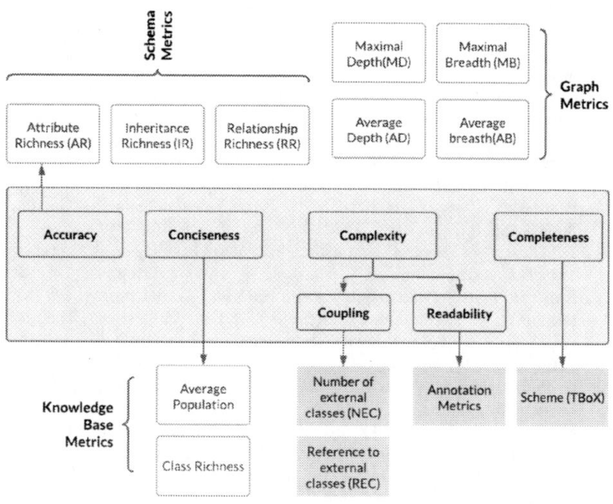

Fig. 1: The metrics used in the evaluation process.

Four criteria are considered: accuracy, conciseness, complexity, and completeness. The accuracy criterion measures the extent to which an ontology models its real-world domain. The schema metrics include Attribute Richness (AR), Inheritance Richness (IR), and Relationship Richness (RR), and graph metrics include Average Depth (AD), Average Breadth (AB), Maximal Depth (MD), and Maximal Breadth (MB). RR indicates the diversity of ontology relationships.

AR indicates the number of attributes (slots) defined for each class, which can be used to infer the quality of the

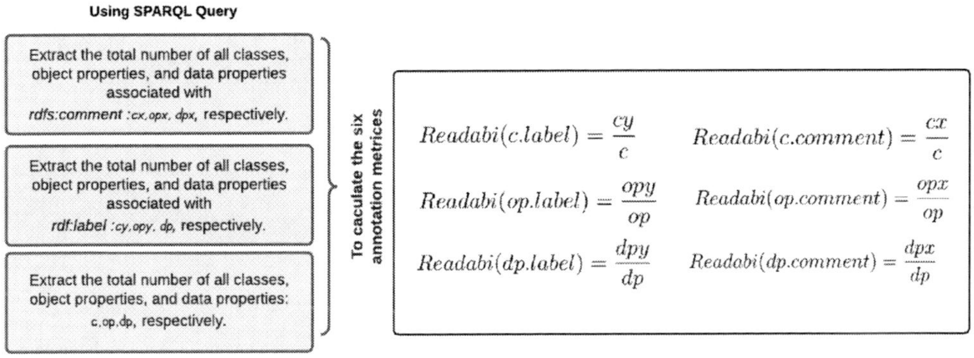

Fig. 2: Steps used to calculate the Annotation Metrics.

ontology design. IR shows the distribution of information across different levels of ontology. The conciseness criterion measures the degree of usefulness of the ontology knowledge. This quality criterion correlates with Average Population (AP) and Class Richness (CR). AP represents the average distribution of instances across all classes. CR is a measure of how instances are distributed among classes. Therefore, it indicates how many instances are related to the classes defined in the schema.

By understanding the complexity of the ontology, developers can better understand, reuse, and reduce maintenance requirements. The coupling and readability of the ontology determined the degree of complexity. Coupling reveals the number of external classes from imported ontologies referenced in the local ontology. It measures the relatedness between the local ontology and other existing ontologies or vocabularies used to construct the ontology [7]. It is defined as:

$$Coupling(O) = REC/NEC \qquad (1)$$

Where NEC is the distinct number of external classes, and REC is the number of references to external classes. The stronger the coupling, the more difficult it is to understand and manipulate. By parsing the OWL file, we calculated the number of distinct external classes defined in the ontology and the number of references to external classes. The code is available online [8].

A measure of readability is the average number of names (labels) and descriptions (comments) per ontology entity, such as classes and properties. We utilize six annotation metrics to evaluate readability. As shown in Figure 2, $c.comment$, $op.comment$, and $dp.comment$ represent the average number of rdfs:comment statements per class, object property, and data property in the ontology, respectively. The average number of rdf:label statements per class, object property, and data property in the ontology is represented by $c.label$, $op.label$, and $dp.label$. GraphDB [6] was used to upload the data sets and to run the SPARQL queries to compute the annotation metrics.

Assessment of completeness considered the schema level and not the instance level of representation. Spatial completeness of the ontologies was done by considering the standard set of possible spatial relationships between data types. For example, there are five possible relationships between two regions, namely, inside, contain, overlap, touch and equal. A completeness score for the ontology is computed in terms of the completeness score of its spatial classes as presented in equation (2), where $Comp$ is the sum of the completeness score of all the spatial classes and C is the total number of spatial classes in the ontology.

$$Completeness = Comp/C \qquad (2)$$

III. Result

Inheritance Richness (IR) shows the distribution of information across different levels of ontology. This metric can distinguish horizontal ontology (where classes have a large number of direct subclasses) from vertical ontology (where classes have a small number of direct subclasses). In Table I, O_2, O_3, and O_4 cover more specific details (depth), while O_1 defines the domain broadly. Relationship Richness (RR) is a measure of diversity of the type of relationships in the ontology. Ontologies O_1, O_2 and O_3 have a low RR score, as they represent mostly one type of relationship; namely the subclass relationship. Result shows that the low value of RR corresponds to higher Maximal Depth (MD) and Maximal Breadth (MB) values. In addition, Attribute Richness (AR) values indicate that the ontologies O_1, O_3 and O_4 contain more attribute information about the classes than O_2.

Average Population (AP) indicates how well the data extraction process was conducted to populate the knowledge base. Results show that O_1, O_2 have a large number of instance per class, indicating a good fit for the class representation in the ontology. CR is related to how instances are distributed across classes. As a result, it displays a percentage indicating the number of instances in each class in the Knowledge Base. The results indicate that O_1, O_2, and O_4 have more instances than O_3. As shown in Table II, O_2 and O_3 have high complexity due to the coupling and readability values. The result indicates

TABLE I: Results of evaluating the ontologies with the graph, knowledge base and schema metrics

Index	Weblink	Classes	Individuals	AR	IR	RR	AD	MD	AB	MB	AP	CR
O_1	https://data.ordnancesurvey.co.uk/ontology/ [accessed: 2023-03-03]	53	2021346	0.321	5.35	0.11	2.38	3	4.33	8	38138.60	0.339
O_2	https://triplydb.com/osi/adminitrative-units [accessed: 2023-03-03]	18	659333	0	0.93	0.166	1.93	2	8	15	36629.61	0.777
O_3	http://linkedopendata.gr/dataset. [accessed: 2023-03-03]	9	2914	0.444	0.88	0.272	1.88	2	4.5	8	323.77	0
O_4	http://data.ign.fr/def/geofla/20190212.en.htm [accessed: 2023-03-03]	8	132567	0.409	0.5	0.56	1.54	3	5.5	15	6025.77	0.409

TABLE II: Evaluation results for the coupling, readability, and completeness criteria.

Index	Coupling	Readability						Completeness
		c.comment	*c.label*	*op.comment*	*op.label*	*dp.comment*	*dp.label*	
O_1	0	0.75	1	0.84	0.84	0.6	0.8	0.56
O_2	15	0.93	1	1	1	0	0	0.3
O_3	8	0	1	0	0	0	1	0.3
O_4	0	0.75	1	0.85	0.85	1	1	0.3

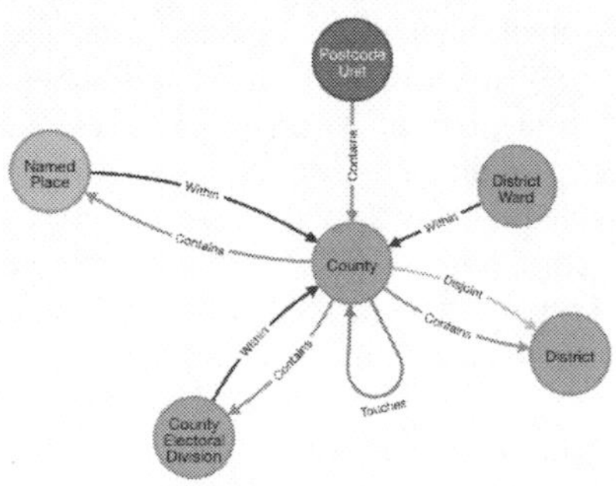

(a) Defined Relationships by OS ontology

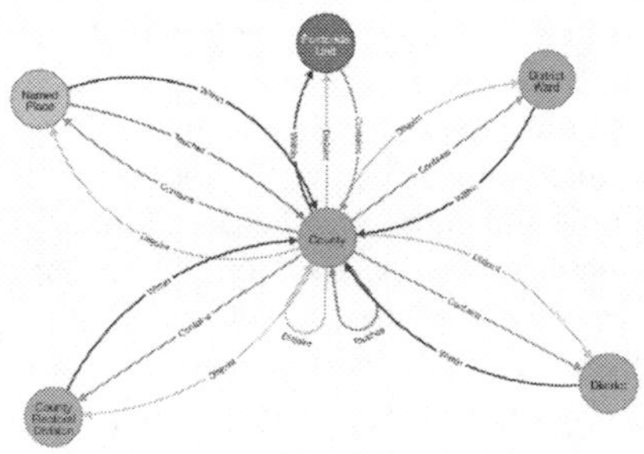

(b) Completed Relationships

Fig. 3: Incompleteness Relationships

that O_2 has a strong coupling, which makes it more difficult to understand and maintain than O_1.

Each class in ontology is checked for completeness, and then equation (1) is used to compute the result.

As an example of incompleteness in the OS is shown in Figure 3, all possible relationships that can be defined between the two polygons, County and District, are $disjoint, contains, within$ as shown in 3b, graph 3a shows the defined relationship between the same two polygons is $within$; thus, the incomplete relationship is disjoint, contains. Results show that O_1 is 56% complete, O_2, O_3, and O_4 are 30%, making O_1 more capable of reasoning and retrieving the geographic information.

IV. Conclusion

In this paper, we evaluated authoritative geographic ontologies using metrics-based methods. Analysis of metrics result indicates that geographic ontologies contain enough data to facilitate knowledge usage. Results confirm that UK ontology covers a wide range of information and show that the ontologies have a good hierarchy. A high score for incomplete spatial relationships leads to fewer inferred geographical details in France, Greece, and Ireland. The UK ontology has very low complexity, which indicates that the model is easy to understand by the user. The future research direction would be to develop a unified data model to integrate the authoritative ontologies.

References

[1] J. Goodwin, C. Dolbear, and G. Hart, "Geographical Linked Data: The administrative geography of Great Britain on the semantic web," Transactions in GIS, vol. 12, pp. 19–30, 2008.

[2] Bucher, et al. "Spatial linkeddata in Europe:Report from spatiallinked data sessionat knowledge graphin action," Official Publication-EuroSDR 73, 2021.

[3] J. Debattista, É. Clinton, and R. Brennan. "Assessing thequality of geospatial linked data–experiences from ordnance surveyireland (osi),", 2018.

[4] M. Poveda-Villalón, A. Gómez-Pérez, and M. C. Suárez-Figueroa, "Oops! (ontology pitfall scanner!)," International Journal on Semantic Web and Information Systems, vol. 10, no. 2, pp. 7–34, 2014.

[5] B. Lantow, "OntoMetrics: Putting metrics into use for ontology evaluation," Proceedings of the 8th International Joint Conference on Knowledge Discovery, Knowledge Engineering and Knowledge Management, 2016.

[6] "Ontotext graphdb," Ontotext, 13-Jan-2023. [Online]. Available: https://www.ontotext.com/products/graphdb/. [Accessed: 04-Mar-2023].

[7] L. Ouyang, B. Zou, M. Qu, and C. Zhang, "A method of ontology evaluation based on coverage, cohesion and Coupling," 2011 Eighth International Conference on Fuzzy Systems and Knowledge Discovery (FSKD), 2011.

[8] H. Muhajab, "HMUHAJAB/coupling," GitHub. [Online]. Available: https://github.com/hmuhajab/Coupling. [Accessed: 06-Mar-2023].

SodaSense: An Intelligence Platform for Micro-Mobility, Micro-Climate and Agriculture

Paraskevi Raftopoulou, Spiros Skiadopoulos, Christos Tryfonopoulos, and Costas Vassilakis
Dept. Informatics and Telecommunications
University of the Peloponnese
Tripoli, Greece
email: {praftop,spiros,trifon,costas}@uop.gr

***Abstract*—SODASENSE is an end-to-end big data manipulation platform that focuses on decision and policy making. It concerns the management of big trajectory and spatio-temporal data that is collected from sensors and relates to micro-mobility, micro-climate and agriculture, and aims to the creation of an expert decision making platform with research, social and business orientation.**

***Keywords*—spatial time series, trajectories, big data management and analysis**

I. INTRODUCTION

SODASENSE (https://sodasense.uop.gr) is an on-going project implemented by the SoDa Lab (https://soda.uop.gr). The platform under development that bears the same name (i.e., SODASENSE) offers end-to-end big data manipulation focusing on decision and policy making in the Peloponnese Region, Greece (https://en.wikipedia.org/wiki/Peloponnese(region)). Specifically, SODASENSE will manage big trajectory and spatio-temporal data that is collected from sensors and relates to micro-mobility, micro-climate and agriculture. An overview of the layered structure of SODASENSE, which comprises of *Data sources*, *Data management* and *Data analysis and visualization*, is illustrated in Figure 1. The uttermost goal of SODASENSE is to create an expert decision making platform with research, social and business orientation

In more details, the platform's services combine the collection, storage, analysis, visualization and data mining, to improve the quality of life in urban centers (through city redesign, planning and micro-transportation) and to support basic economic activities (such as tourism and agriculture), especially for small and medium-sized enterprises that make up most of the economic backbone of the Peloponnese. The SODASENSE project is designed to enhance with *specific services* the coordination of public authorities and the relevant policy decisions making in the areas in which it specializes. Through the new services, it will provide its users with information, knowledge and tools on:

- the use of new technologies in micro-mobility, micro-climate and agriculture,
- the collection of relevant data, and
- the utilization of data to improve life in urban centers.

Overall, SODASENSE will offer additional capabilities and services to academia, research, professional associations, private and public companies, infrastructure operators, independent authorities and public bodies, to support original research, investment projects and decision making. This will enhance tourism, support agricultural production, and better facilitate decision-making in the Peloponnese area.

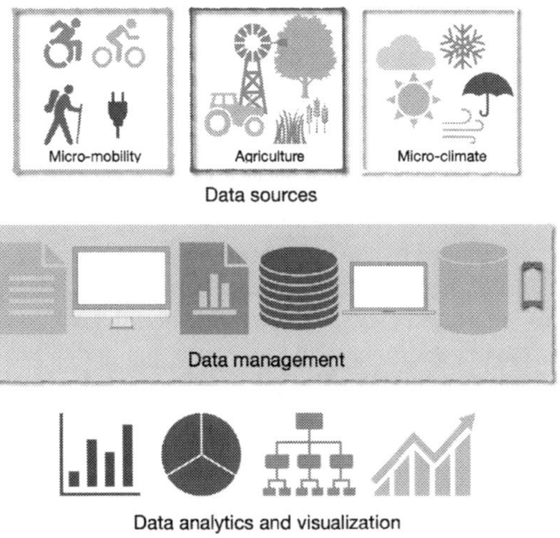

Figure 1. SODASENSE in a nutshell

The rest of the paper is structure as follows. Sections II and III illustrated the added value and the approach of SODASENSE. Finally Section IV offers conclusions.

II. SODASENSE ADDED VALUE

Greece, due to its geographical position and its history, is a pole of attraction for tourism. Consequently, there is a great demand for micro-movements of both tourists and permanent residents of urban areas. At the same time, it has a long tradition on agriculture and the cultivation of excellent quality agricultural products (e.g., olives and olive oil, grapes, raisins and wine, figs). Until the launch of the SODASENSE infrastructure, no other infrastructure had been created for the systematic research, analysis and dissemination of data related to micro-mobility, micro-climate and agriculture.

The lack of infrastructure to provide information and data on the above subjects entails additional constraints for conducting further research. It also limits the development of strategic plans and policy making at local, regional and national

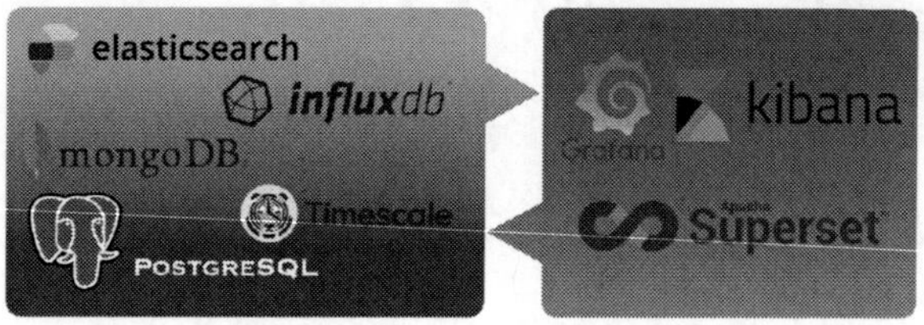

Figure 2. SODASENSE technologies

level. The SODASENSE platform responds to the above needs by updating and monitoring data on micro-mobility, micro-climate and agriculture, combining them with data from other stations (where available), as well as identifying and recording trends with a focus on smart specialization in these areas. The specialized knowledge that SODASENSE provides will also enhance the implementation in Greece of European directives and policies related to the aforementioned areas.

Overall, the main objective of the SODASENSE infrastructure is to enhance regional excellence in the fields of micro-mobility data management, micro-climate and agriculture, by providing the necessary services (data collection and analysis, modeling, programming, applications to users).

III. THE SODASENSE APPROACH

SODASENSE platform will provide an reliable and easy access to data using information technology and communication services. SODASENSE addresses innovative topics, such as large-scale data management, real-time exchange of scientific data, data processing with advanced artificial intelligence and machine learning methods, the use of cloud computing services, aiming to improve the efficiency and reliability of activities critical for a modern urban environment, such as micro-mobility, micro-climate, smart agriculture, as well as tourism.

More specifically, a Data Center will be designed as an application of the Database as Service (DaaS) model [1]. SODASENSE's services will be provided under the Software as a Service (SaaS) model [2], while the Services Center will operate under the Internet as a Service (IaaS) [3] and Platform as a Service (PaaS) models [4]. Figure 2 summarizes the technologies and tools employed by SODASENSE.

In the field of micro-mobility, micro-climate and agriculture, the scientific and technological innovation of the operation lies in the use of digital methods to gain new knowledge about large-scale data, advanced computational models, applications and visualization technologies. Research will be carried out on micro-mobility topics, such as routing, compliance with the new regulations regarding the environmental impact of micro-mobility, intermodality and accessibility to poles of concentration of cultural or educational activities.

Accordingly, urban air quality simulations will be designed and carried out taking into account buildings of different shapes, sizes and layouts. Finally, research will be carried out on the placement of smart sensors on crops, as well as combining them with data from other stations where available, in order to collect data that will help farmers monitor and optimize crops in line with changing environmental factors. At the academic level, the scientific innovation of practice lies in the use of advanced methodologies and tools with increased reliability, such as simulation, decision-making and policy support models.

SODASENSE aims to become an innovative e-research infrastructure with significant added value at all levels: regional, national and international. At regional level, it will support smart specialization, assisting regional innovation and economic development of the Peloponnese Region, allowing regional resources to focus on their advantages, including the wealth and smart specialization of small and medium-sized enterprises, and in particular small and medium-sized enterprises engaged in micro-mobility, agriculture and tourism.

IV. CONCLUSIONS

SODASENSE infrastructure will support in an interdisciplinary way the key axes pertaining to micro-transport, micro-climate, agriculture and tourism. It will create an electronic information system for the systematic information, decision support and research, while the use of standards for the interoperability of the collected data and the provision of open availability of the data will allow the exploitation of datasets and related tools by the research potential and the general public.

Finally, SODASENSE is expected to have a positive impact on various socio-economic sectors specific to the region in the context of smart specialization (e.g., tourism, micro-mobility, micro-climate, environment, agriculture). At a different level, through SODASENSE the region will have, for the first time, an integrated tool (and the relevant data) to study and support decision-making of key areas of activity; the platform has the potential to promote green micro-transport, precision agriculture, and sustainable tourism development. From the above, it is understood that there is a wide range of beneficiaries of the operation at both national and regional level, including regional and local authorities, organizations, businesses and bodies in the target sectors.

ACKNOWLEDGMENTS

This research was funded by project SODASENSE (https://sodasense.uop.gr) under grant agreement No. MIS 5060275 (co-financed by Greece and the EU through the European Regional Development Fund).

REFERENCES

[1] H. Hacigumus, B. Iyer, and S. Mehrotra, "Providing database as a service," in *Proceedings of ICDE*, 2002, pp. 29–38.
[2] B. Allen et al., "Software as a service for data scientists," *Commun. of ACM*, vol. 55, no. 2, p. 81–88, Feb 2012.
[3] B. Wang, X. Chang, and J. Liu, "Modeling heterogeneous virtual machines on IAAS data centers," *IEEE Comm. Letters*, vol. 19, no. 4, pp. 537–540, 2015.
[4] R. Dua, A. R. Raja, and D. Kakadia, "Virtualization vs containerization to support PAAS," in *IEEE Int'l Conf. on Cloud Engineering*, 2014, pp. 610–614.

The Role of Food Sharing Applications (FSAs) in Supporting Food Waste Mitigation in Cities – Examples from Poland and Czechia

Magdalena Pączek, Mateusz Rusin, Filip Bolicki, Daniel Straub, Magdalena Kubal-Czerwińska
Institute of Geography and Spatial Management
Jagiellonian University, Kraków, Poland
email: magda.paczek@student.uj.edu.pl, mateusz.rusin@student.uj.edu.pl, filip.bolicki@student.uj.edu.pl, d.straub@doctoral.uj.edu.pl, magdalena.kubal@uj.edu.pl

Abstract— **The current problem of reducing food waste is becoming the subject of frequent political debate in the international arena. Researchers of global changes in the natural environment, as well as representatives of national, local, and non-governmental organizations, realize, as part of their activities, that a large amount of food cannot be saved. This increased attention is not surprising, as food is a basic need. Online platforms and mobile applications contribute to the redistribution of food saved thanks to modern information and communication technologies. Online platforms and application connect local entrepreneurs and the catering sector with their customers directly, changing the way food distribution works. Our research includes a food sharing application (FSA) as one of the practical solutions, in the business to customer (B2C) model, which directly connects catering or food retail facilities with the customer, supporting food waste mitigation in cities, on the example of FSAs used in Poland and Czechia. Based on data on localization and the type of food saved in specific areas of selected cities, it was possible to determine the regularities of the spatial distribution of food places and their relation to socio-demographic characteristics. Using their localization pattern, we were able to draw a picture of the food saving opportunities of catering facilities, operating through the B2C (business to customer) marketing model. Additionally, attitudes and opinions about joining FSAs, will be discussed along with the benefits and disadvantages of participating in similar initiatives.**

Keywords-food sharing applications (FSA); food waste; food waste mitigation; Poland; Czechia.

I. INTRODUCTION

Today, the problem of reducing food waste is becoming the subject of frequent political debate in the international arena. Researchers of global changes in the natural environment resulting from the waste of food products, as well as representatives of national, local governmental and nongovernmental organizations, realize, as part of their activities, that a large amount of food wasted can no longer be saved. This increased attention is not surprising, Despite being a basic need food is wasted in large quantities at all stages of the food supply chain [1]. Local initiatives to collect, manage, and share surplus food have been facilitated by the development of emerging information and communication technology, such as online platforms and food sharing applications (FSA). Although food sharing is often discussed as a potential improvement to the food system, empirical research is still scarce, and few researchers have explored the reasons why practitioners, including trade and service entities, join food sharing initiatives. Food sharing initiatives are often considered transformative at various scales (national and regional food banks, church organizations, or small local eateries).

Geospatial information systems support many aspects of geospatial data collection, presentation, processing, visualization, sharing, and management, and provide information on many geo-environmental hazards and how to manage them sustainably. GIS (Geographic Information System) environments today enable food waste management and modelling [2][8][9] by providing actors focused on food waste management tools for mapping food waste locations based on waste types, waste quantities, and other variables, which will improve the management of this problem on a daily basis and in the long run [3][4]. A database or GIS system can provide information on waste locations on large scales ranging from general city maps to detailed maps of food waste in specific areas [5]. By using GIS tools on a regional to national scale, it is possible to combat food insecurity and promote social justice [6][7][11]. GIS (Geographic Information System) tools, such as geotagging, geofencing, cluster analysis, and geoprocessing frameworks, can solve the social and environmental problems of food waste [9][10]. Through GIS (Geographic Information System) environments, food waste application developers assist in localizing specific entities that offer food at a reduced price, while users of those applications can see what food they can purchase and where it is available, helping to minimize food waste.

With the development of modern information and communication technologies, such as online platforms and mobile applications, food is being redistributed online by connecting local entrepreneurs and the catering sector with their customers directly, e.g., TooGoodtoGo (the United Kingdom, Denmark, Germany, Poland), Foodsi (Poland), Karma, (the United Kingdom), Nesnezeno (Czechia), Olio (the United Kingdom), as well as donors with charities (Italy: BringTheFood, the United Kingdom: Olio). FSAs, whether profit and non-profit orientated [12], change the way food distribution works, from linear (production to supermarket/restaurant to consumer) to network (e.g., customer to producer) [13]. In the market, there are already various digital platforms for reducing food waste and sharing food (internet platforms, mobile food applications) and their essential relates, among other, to communication [14].

Our research includes food sharing applications as one of the practical solutions, in the B2C model, to food waste. Food sharing applications communicate about available food for consumption, directly connecting catering facilities or food retail facilities with the customer, supporting food waste mitigation in cities, on the example of such applications from Poland and Czechia.

II. RESEARCH QUESTIONS

In our analysis of the FSAs used by gastronomic facilities (e.g., restaurants, cafes, etc.) and food stores, located in selected cities of Poland and Czechia, the following research questions emerged:

- What is the access to food from the apps in the selected cities in Poland and Czechia?
- What kind of food is saved in those cities?
- What are the regularities of the spatial distribution of gastronomic facilities in the selected cities (and the connections with the socio-demographic aspects and functions of the city)?
- What picture of food-saving opportunities in the city emerges from their spatial distribution?
- Are there city districts where the food cannot be saved from waste through the use of food apps?

III. METHODOLOGY

To provide the answer to these research questions, two-step data collection methods were implemented. In the first stage, the investigators, trained to prepare a required database, collected data for analysis using three FSAs: Foodsi and ToGoodToGo for Poland; Nesnezeno for Czechia. The data was collected between September 2022 and January 2023. The investigators collected 1246 records corresponding to broadly understood facilities from the catering industry (e.g., restaurants, hotels, cafes, bars, etc.) and establishments related to the food trade (gas stations, grocery stores, bakeries, confectioneries, etc.) from 4 urban regions in Poland: Warsaw, Cracow, Poznan and cities that are part of the Upper Silesian-Zaglebie Metropolis and 2 cities from the Chechia: Prague and Brno. All the records collected from selected urban regions were merged into one database. During this process, some facilities repeated in the database because they were present in both Polish FSAs. This was the case with 51 records, The duplicates were removed from the final dataset. This means that the final database contains in total 1195 records. Regarding the study area, these are all large urban centres of a multifunctional character, to which Poles and Czechs migrate, among others, to study and remain, supplying the labour market. These are well-known international centers, cosmopolitan centers with the rank of European cities, whose inhabitants are usually characterized by higher education and employment in the service sector. Warsaw, Cracow, Prague, and Brno are also well-known urban centers, historical cities with long tourist traditions, visited year by year by an increasing number of domestic and international tourists. A large group of residents of these urban centres are also expats who come to Warsaw, Cracow, or Prague to join the ranks of employees of international corporations [15; 16]. The collected databases were combined for further quantitative and qualitative analyses. In the data collection process, special attention was paid to obtain information on the location of catering facilities and grocery stores, the type of facility, the type of food saved from waste, the characteristics of the facility in terms of ownership, etc.

As part of the second stage of research, qualitative data will be augmented with quantitative data to identify attitudes and opinions among owners of catering facilities and food entrepreneurs. In addition to the benefits and disadvantages of participating in similar initiatives, we discussed attitudes and opinions about joining the food protection program through food sharing applications.

IV. RESULTS

The current stage of research ended with the acquisition of information on the location and characteristics of facilities participating in food-saving activities by selling products that are past their best-before date directly to customers using mobile application channels. Based on data on localization and the type of food saved in specific areas of selected cities, it was possible to determine the regularities of the spatial distribution of food places and their relation to sociodemographic characteristics. Using their localization pattern, we were able to paint a picture of the food-saving opportunities of catering facilities operating through the B2C marketing model. The most numerous restaurant facilities in the food sales program that use the FSA are those providing services for the preparation and sale of ready meals, such as restaurants, cafes, and buffet hotels. Another group of facilities are shops, including neighborhood greengrocers, bakeries, and large-area stores (so-called supermarkets). Most often, facilities that sell food through the FSAs are in central districts (e.g., Cracow, Warsaw, Prague, Brno) with diverse functions (e.g., tourist, administrative, entertainment). There are also facilities joining the FSA in typically residential areas, often revitalized,: e.g., in Warsaw, Cracow, Poznan, or Brno (Fig. 1)

Prague is a special case, since FSA-using facilities are concentrated in the city's central tourist district, and their occurrence decreases with distance from the centre to residential areas (Fig. 2).

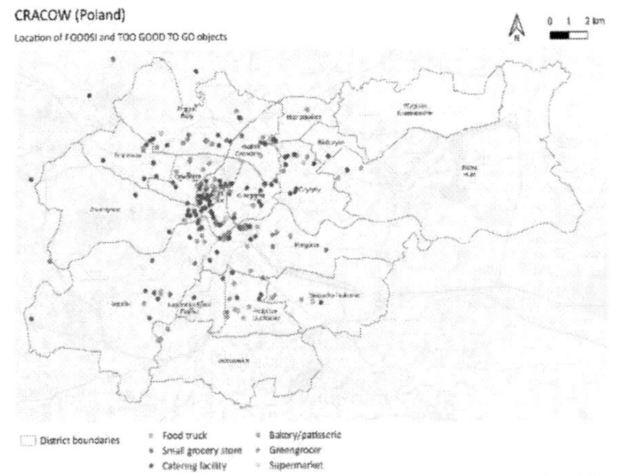

Figure 1. Distribution of gastronomic premises that save food from waste using FSAs in Cracow (Poland)
Source: own elaboration.

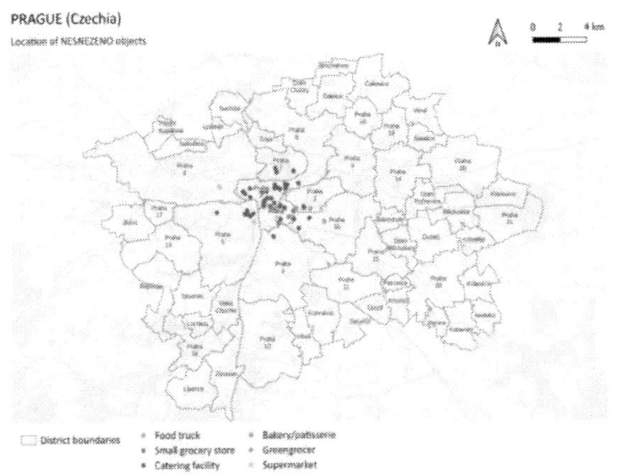

Figure 2. Distribution of gastronomic premises that save food from waste using FSAs in Prague (Czech Republic)
Source: own elaboration.

V. Conclusion and Future directions

The use of data from mobile applications dealing with the reduction of food waste from catering facilities allowed some preliminary conclusions. Both in the central districts of large cities, with various metropolitan functions, and in districts far from the center (mainly residential), there are facilities that use FASs to reduce food waste. Residents of both types of districts are, therefore, equipped with tools that allow them to participate in this effort to reduce food waste. It might be assumed that the owners of the premises, located in those areas, represent pro-environmental attitudes to food waste and that the young residents of these districts, Millennials and Generation Y, are the main users of those services. Since the above-mentioned services and products sell quickly, it is assumed that there is a demand for them in these city districts, both among sellers and consumers.

In the future, in addition to spatial analysis of the distribution of entities joining FSAs in the selected cities in Poland and Czechia, the benefits and disadvantages of participating in similar initiatives, as well as attitudes and opinions about joining FSAs, will be discussed. Participants in food sharing have a variety of motivations and objectives that can mutually strengthen each other. We are interested in determining whether entities can simultaneously participate out of moral imperative and with a desire to make quick money. However, there can be tensions between individuals participating in collective action because of their diverse motivations and expectations. Some participants see their participation as a reflection of certain values (morality), while others see it primarily as an opportunity for personal gain (for example, by using the application as an additional distribution and sales channel). From the perspective of the methods introduced, the analysis pinpoints the need to search for unified and general motives of the owners of such premises to help residents of neighbourhoods participate in food waste mitigation practices, using phone applications.

Open questions remain about the discrepancies between different individual views on what food sharing should and can achieve using FSAs. From the point of view of new owners, for whom both values: pro-environmental attitudes and environmental responsibility are important, it is extremely important crucial to determine what motivations drive the owners of catering facilities and entrepreneurs when joining the above-mentioned applications. Also, as documented, gastronomic facilities tend to appear in clusters, therefore, the therefore question of whether joining FSAs is subject to policy diffusion remains to be addressed.

Acknowledgment

This research is part of the Twinning project "Building excellence in research of human-environmental systems with geospatial and Earth observation technologies" that has received funding from the European Union's Horizon 2020 research and innovation programme under grant agreement No 952327.

References

[1] A. S. Stenmarck, C. Jensen, T. Quested, G. Moates, M. Buksti, B. Z. Cseh, D. Juul, A. Parry, A. Politano, and B. Redlingshofer, "Estimates of European Food Waste Levels". IVL Swedish Environmental Research Institute, 1996.

[2] H. V. Raut, A. D. Nagne, and K. V. Kale, "Management of food waste by using GIS environment: a review", International Journal of Computer Sciences and Engineering, vol 3., issue 7, pp. 144-151, 2015.

[3] S. Karsauliya, "Application of remote sensing and GIS in solid waste management: a case study of surroundings of River Yamuna, India", International Journal of Environmental Engineering and Management, , Vol. 4, pp. 593-604, 2013.

[4] Draper/Lennon, Inc. "Identification, characterization, and mapping of food waste and food waste generators in Massachusetts", http://www.mass.gov/eea/docs/dep/recycle/priorities/foodwast .pdf, 2002, September.

[5] "Geographic Information Systems for Organics Recycling", BioCycle, https://www.biocycle.net/2010/03/23/geographic-informationsystems-for-organics-recycling/, 2010, March.

[6] J. Baumann, "GIS project helps with food insecurity and waste", https://www.esri.com/about/newsroom/wp-content/uploads/2020/11/foodinsecurity.pdf, 2020.

[7] L. Mabe, "Food Waste Geographies: a GIS-based spatial analysis of food waste in Los Angeles County" UC Davis, ProQuest ID: Mabe_ucdavis_0029M_20464, 2021, ark:/13030/m5rc439s.

[8] A. Singh, "Remote sensing and GIS applications for municipal waste management", Journal of Environmental Management, Vol. 243, pp. 22-29, 2019, https://doi.org/10.1016/j.jenvman.2019.05.017.

[9] E. Mati Asefa, K., Bayu Barasa, and D. Adare Mengistu, "Application of Geographic Information System in solid waste management" IntechOpen, 2022, doi: 10.5772/intechopen.103773.

[10] S. K., Gurjar and A. Gaur, "21 - Application of remote sensing and GIS in integrated solid waste management - a short review", Editor(s): Ch. Hussain, S. Hait, Advanced Organic Waste Management, Elsevier, 2022, pp. 351-362, https://doi.org/10.1016/B978-0-323-85792-5.00001-0.

[11] M. Mathenge, B. G. J. S. Sonneveld, and J. E. W. Broerse, "Mapping the spatial dimension of food insecurity using GIS-based indicators: A case of Western Kenya" Food Security, vol. 15, pp. 243–260, 2023, https://doi.org/10.1007/s12571-022-01308-6.

[12] J. Harvey, A. Smith, and D. Golightly, "Giving and sharing in the computer-mediated economy", Journal of Consumer Behaviour, vol. 16, pp. 363– 371, 2016, doi: 10.1002/cb.1499.

[13] J., Harvey, A. Smith, J. Goulding and I. B. Illodo, "Food sharing, redistribution, and waste reduction via mobile applications: A social network analysis", Industrial Marketing Management, vol. 88, pp. 437-448, 2020, https://doi.org/10.1016/j.indmarman.2019.02.019.

[14] A. Rýparová, "Digital food sharing: motivation and practice of sharing in Czechia" Geografie, vol. 126, pp. 263-287, 2021, https://doi.org/10.37040/geografie2021126030263.

[15] National Statistical Office. "Narodowy Spis Powszechny Ludności i Mieszkań 2021 [National Population and Housing Census 2021]", Polish Statistical Office, 2022.

[16] Czech Statistical Office, "Statistical Yearbook of Prague – 2022", Czech Statistical Office, 2022.

International Academy, Research,
and Industry Association (IARIA)
412 Derby Way
Wilmington, DE 19810

ISBN 978-1-7138-8060-8

CURRAN ASSOCIATES INC.
proceedings.com

Third International Conference on Advances in Sensors, Actuators, Metering and Sensing (ALLSENSORS 2018)

Held at DigitalWorld 2018

Rome, Italy
25 - 29 March 2018

IARIA